AF541366

ps
Pvt. Ltd.

Parthenium

Insight into its Menace and Management

M. Mahadevappa

Director, JSS Rural Dev. Foundation,
Mysore 570 004
and
Ex Chairman, ASRB, *Ex* Vice-Chancellor UAS,
Dharwad # 1576, 1st Cross, Chandra Layout
Bangalore 560 040, Karnataka

2009

Studium Press (India) Pvt. Ltd.

Parthenium
Insight into its Menace and Management

ISBN: 978-93-80012-13-1

Published by:

Studium Press (India) Pvt. Ltd.
4735 / 22, 2nd Floor, Prakash Deep Building
(Near Delhi Medical Association),
Ansari Road, Darya Ganj, New Delhi-110 002
Tel.: 23240257, 65150447; Fax: 91-11-23240273;
jngovil@gmail.com; jngovil@hotmail.com

Printed at:

Salasar Imaging Systems
Delhi-110035 (India)

FOREWORD

Parthenium, an obnoxious annual weed, has been reported as a main source of health hazard to mankind and animals, threat to biodiversity and danger to environment. Since its introduction into India in 1956 through grain shipment, it has spread alarmingly like a wild blaze in almost all the states, It has also invaded forest areas, pastures and garden lands.

Parthenium is a highly regenerative plant, insensitive to both photo-and thermo period and possesses remarkable adaptability to all agro-ecological habitats. It is threatening both social environment and biodiversity due to its adverse effects on man, domestic animals and beneficial flora. Therefore, concerted peoples' participatory approach is critical for its effective control.

It is encouraging that scientists have done considerable research on various methods of its control. By and large, these methods do have in-built limitations. Therefore, an integrated approach becomes an obvious choice for its sustainable management. The literature in this regard, besides being scanty, has been widely scattered. Hence, there is an urgent need to have consolidated information in the form of a book, which is being fulfilled due to efforts of Padmashri Dr. M. Mahadevappa, former Chairman, ASRB, New Delhi and Vice-Chancellor, University of Agricultural Sciences, Dharwad, Karnataka, India.

Dr. M. Mahadevappa has been associated with the problems and issues of this weed for nearly two decades. He has evolved strategies to contain this weed through his associates, voluntary organizations and civic authorities in Karnataka. He had also organized two International Conferences on this subject in the State of Karnataka where the menace of this weed is considerably brought down through integrated pest management.

I do hope that Dr. Mahadevappa, with his vast experience in this field and his current association with NGOs, will continue to help combating Parthenium problem through research, extension and publications of this kind. I am confident that this publication will prove very useful for agricultural scientists, doctors, environmentalists, planners and common man.

Dr. R.S. Paroda
Chairman
Trust for Advancement of Agricultural Sciences
IARI Campus, New Delhi
&
Former Secretary DARE
and Director General, ICAR

Date: 7th August, 2009

ABOUT THE AUTHOR

Padmashri Mahadevappa (Rice Mahadevappa), through his sustained breeding efforts at UAS Bangalore and at IRRI developed and released nine rice HYVs/hybrids contributing to nation's exchequer substantially. He developed biological eco-friendly Integrated Parthenium Weed Management (IPWM) Technology. He visited 21 nations and presented 31 papers, published 200 research articles, besides authoring 17 books on rice, Seed Technology and Parthenium in English and Kannada. As Vice Chancellor of UAS, Dharwad for two terms, he contributed immensely to get ICAR's "SARDAR PATEL OUSTANDING INSTITUTION AWARD FOR 2000". As chairman ASRB, he brought transparency in recruitment and expedited the process of academic promotions and encouraged scientists to further contribute to the cause of farming community.

Several reputed organizations have recognized Professor Mahadevappa's rich contributions for the cause of agriculture and the farming community. He was conferred the coveted (International) Watumull Foundation Award (1987), Hooker

Award (1981) and Sir Chotturam National Award. The Karnataka State conferred the Rajyotsava Award in 1984. In addition he was conferred the Nagamma Dattathreya Award by UAS, Bangalore, and Parisara Ratna Award and Bharatha Rathna Sir M. Visvesvaraya Memorial Award (1999). It is in 2005, Dr. Mahadevappa was recognized by President of India by awarding "Padmashri".

PREFACE

The problem of Parthenium weed has been apparent since more than five decades and is continuing to affect health of humans, animals, flora and environment. It distorts the aesthetic look of the natural as well as manmade landscapes. It is considered as one of the most noxious terrestrial weeds in India. Today, probably more people are aware of its obnoxious effects and thinking of principal issues to arrest its menace, if not eradicate it immediately. Principally, no plant should be eradicated, as nature's every creation has some role to play some time and somewhere on this planet; hence it is important to see that its harmful effects to living beings and environment are managed.

Further, the limited success so far achieved by scientists and various societal communities to manage this weed indicates that it needs a thorough analysis of the major obstacles that are being faced and likely to be so in future too. The factors involved are too many and it calls for deeper understanding for an integrated approach to design appropriate and sustainable management tactics.

The current effort therefore embodies aspects of what has happened and what will happen if neglected and what needs to be done after understanding its present status. The unpredictable nature of the factors afflicting the weed compels us to debate the concerns, our capacity to manage, the resources, allied and current technical knowledge in the light of our achievements and success stories with modicum of facilities.

It is in this context this book is prepared and it is hoped that this book provides an insight into the special features of the plant and its very wide adaptability, overall strategies for developing a viable and workable policy to contain the weed and minimize its negative effects. It also deals with potentials of government assistance, research and extension management, need for involvement of civil societies and also assessment of the economically feasible options.

Another objective of the book is to collate available and relevant technical and non technical information and to bring to the notice of scientists, students, public and administrators as how to minimize the harmful effects of Parthenium. Hopefully, this endeavor may lead to more enlightening efforts in future to make our environment free from its ill effects and throw light on how to harness possible benefit from this rare kind of plant. The book has seven major chapters dealing with 1) The plant and its ecology, 2) Use and misuse of Parthenium, 3) Spread of Parthenium, 4) Harmful effects of Parthenium, 5) Methods of control, 6) Weed control act, 7) Coordination of control programme and a few compilations on International Conferences, success stories, scope for further research as well as FAQs.

Further, this publication is aimed to put together all the information about the pernicious weed Parthenium in a simple language, to be comprehensive and to induce interest among students, scientists, researchers, teachers, progressive farmers, municipal authorities, industry people and all those who have keen interest in conserving the natural flora and fauna as well as those having consciousness to protect the environment.

ACKNOWLEDGEMENT

The multitude of activities in vogue has prompted me to write this comprehensive book for the benefit of all involved in maintaining environmental cleanliness. In this endeavor I was conscious of the assiduous efforts made by scientists, environmentalists, policy makers and grass root level workers towards achieving environment security by reducing the ill effects of the weed Parthenium.

I place on record the hard work, commitments and instructing efforts put by my colleagues and friends in preparation of this book. Several organizations, institutions and individuals have helped me during the preparation of the book. I thank all of them.

I wish to record my special thanks to Dr. Pankaj Odhia, IRPNG, Raipur who has dedicated himself for parthenium related activities across the globe and whose website has provided me lot of information that is covered in this book. I am greatly indebted to Dr. K.V.S.R. Kameswar Row for going through the manuscript repeatedly and suggesting modifications. The help of Dr. T.M. Manjunath, Monsanto, Bangalore and Dr. Ramachandra Prasad, Professor of Agronomy (Weed Science), UAS, Bangalore and Dr. R.D. Gautam, Professor of Entomology, IARI, New Delhi for going through some chapters and giving suggestions. I am thankful to my wife Smt. Sudha Mahadevappa for cooperating in many ways while preparing this book. I am extremely grateful to Miss. Yogeshwari C. for making type script from the beginning to end in the preparation of this book.

M. Mahadevappa

CONTENTS

LIST OF TABLES

LIST OF PLATES

One

INTRODUCTION

Parthenium (*Parthenium hysterophorus* L.) a waste land weed, is a herbaceous, erect and annual plant belonging to the family Asteraceae. It has several synonyms like *fever few*, *congress weed*, *star weed, carrot weed, white cap, white top, naxalite weed* etc. and in Indian languages it is known as *chatak chandani*, *broom bush*, *osadi, gajari-kasa, phandriphulli, nakshathra gida, vyari bama and safed topi*. Centuries ago it was noticed in North and South America, Cuba, West Indies, Vietnam and was reported from Australia, Taiwan, Southern China, Pakistan, Bangaldesh, Pacific Islands, East and South Africa and Canada.

The origin of this obnoxious weed (*Parthenium hysterophorus)* dates back to 4th Century BC in the Greek city of Parthenia. This weed has been named so according to the name of the Greek City. It was distributed throughout the world along with wheat. The country which adopted wheat as its staple food, that country was affected by the effects of this Parthenium. Its adverse effects are reported largely in African, Australian and Asian countries. Other sister species occur in the European countries and their names are given elsewhere in this compilation. Even though it did not find any place in the world's worst weeds till 1977 (Holm *et al*., 1977), in the last few decades it has become one of the seven most dreaded weeds of the world (Singla, 1992). On the basis of information on occurrence of Parthenium given by Towers *et al.* (1977), the world distribution of Parthenium along with distribution in India was presented by Aneja *et al.* (1991). After noticed in mid-fifties, its spread throughout the country was found very rapid and of abnormal density; affecting the life of man, livestock and other plant species. The weed is presumed to have accidentally crept into Maharashtra along with the food grains imported from the United States of America (Vartak, 1968). It is also opined that the seeds of this weed were accidentally introduced with the cereals imported from USA and cultivated experimentally at the Pune Agricultural College Farm (Lonkar *et al.,* 1974). However, some reports trace its history of occurrence to about one century back (Roxburghi, 1914; Maiti, 1983; Dam *et al.,* 1993). Its presence in India before 1955 got further confirmation from a herbarium record in Forest Research Institute, collected by Dr. Brandis in 1880 (Bennet *et al.*, 1978). From these records, it is clear that *P. hysterophorus* had entered in India before the start of 20th century and survived not so visibly till 1955. Why it did not spread despite its most favorable features to be invasive is a point that creates doubt whether it existed at all in India. But it is true, that, since 1955, Parthenium has spread like a wild fire throughout India. Dissemination of the seed is attributed to imported cereals from outside, as the seeds of parthenium which are tiny and well structured can cling to any play-material, suitcases, clothes, shoes etc. and can be dispersed.

Several factors such as (i) the absence of natural agents that restrict the vigour and aggressiveness of this plant as in its original home, (ii) high fecundity, (iii) presence of efficient seed dispersal mechanism it is possessing, (iv) monopolistic allelopathic impact on most other plant species, (v) unsuitability for grazing because of the presence of anti-feedants in the plant system and

(vi) efficient biological activity and wide adaptability to varying soil, agro-climatic and micro environmental conditions, have enabled this plant to invade a variety of growing environments particularly in situations associated with human activities. The Parthenium seeds can germinate, flower and set seeds within four weeks under favorable conditions. Once it gets established, it can survive severe droughts and frosts. Thus, all human efforts to control it have met with little success. Hence, management of this weed has posed a serious challenge to the scientists, administrators and the public.

In India, Parthenium has spread gradually from one place to another and is now commonly seen all along the highways, petrol bunks, railway tracks, road sides and other waste lands including burial grounds. It can establish in neglected, over stacked areas and also in disturbed, or degraded or bare soils. Incidentally, it has also found its way into agricultural fields through city wastes lifted by the farmers. It has also spread through rain as well as sewage water in the residential and industrial areas, vacant sites and other fallow lands and abandoned fields.

The consequences of such rapid and gregarious growth of Parthenium, especially its ill-effects on human, livestock and plants, have been well published to create public awareness. Manifold – manual, chemical, cultural and biological approaches have long been attempted to contain this weed, but positive results are seen only through integrated management practices suggested from 1988 onwards, both in India and Australia. Information on the structure and distribution of Parthenium, its behavior, and impact of its spread on the social environment, methods of its control, merits and limitations of recommended methods to control are discussed in this publication.

Two

THE PLANT AND ITS ECOLOGY

Botanical Characteristics

The genus *Parthenium* consists of 23 species, spread over the western hemisphere and five of them occur in Mexico itself (Hakoo, 1963). While the basic chromosomal number for this genus is x=9 (Darlington and Janaki Ammal, 1945), the chromosome number of *P. hysterophorus* has been reported by Hakoo (1963) to be diploid, 2n=18. All of them belong to the family *Asteraceae* (*Compositae*). The weed has botanical affinity with the species of genus *Ambrosiam* (Castex *et al.*, 1940). The genus according to weed scientists, has the following species:

[i] *Parthenium alpinum*

[ii] *Parthenium argentatum*

[iii] *Parthenium auriculatum*

[iv] *Parthenium bipinatifidum*

[v] *Parthenium cinerace*

[vi] *Parthenium confertum*

[vii] *Parthenium densipilum*

[viii] *Parthenium divericatum*

[ix] *Parthenium fruticosum*

[x] *Parthenium glomeratum*

[xi] *Parthenium hysterophorus*
[xii] *Parthenium incacum*
[xiii] *Parthenium integrifolium*
[xiv] *Parthenium ligulatum*
[xv] *Parthenium lozanianum*
[xvi] *Parthenium lyratum*
[xvii] *Parthenium microcephalum*
[xviii] *Parthenium rollinsianum*
[xix] *Parthenium schotti*
[xx] *Parthenium stramonium*
[xxi] *Parthenium tetraneuris*
[xxii] *Parthenium tomentosum*
[xxiii] *Parthenium trilobatum*

The morphology and some important characteristics of Parthenium are detailed below:

Systematics and eco-morphological characters

Character	Description
Genus, Latin	*Parthenium*
Species, Latin	*hysterophorus*
Author	L.
Family	COMPOSITAE (ASTERACEAE)
Common name, English	**Santa-Maria**
Chorotype	American
Life form Raunkiaer	Therophyte, annual
Summer shedding	Ephemeral
Succulence	**Non-succulent**
Salt resistance	Glycophyte
Habitat or affinity to vegetation formation	Disturbed ground at roadside managed with herbicides, irrigated deep clayey soils of the valleys in the northern parts of the country.

Scientific name	:	*Parthenium hysterophorus* L.
Sub tribe	:	Ambrosiinae
Tribe	:	Heliantheae
Family	:	Asteraceae
Common name	:	Parthenium weed, Parthenium, white top

This annual herb grows to a height of 0.5 to 1.5 m and very rarely up to 2.0 m with a tendency to attain perennial habit (Plate 1). The stem is angular and longitudinally grooved, profusely branched and highly pubescent with short whitish hairs visible only under close observation. The initially emerged leaves enlarge gradually to form a rosette (Plate 2) close to the soil surface. The leaves which develop later emerge from the extended flowering branches which are relatively small and are placed distantly compared to the first formed rosettes. The leaves are alternate, sessile, irregularly dissected and bipinnate, resembling that of carrot or chrysanthemum and have tiny hairs on both the surfaces. The leaves are deeply lobed light to dark green. The plant has dense pubescent soft hairs. When the plants rub each other or rattle due to wind action, the hairs break off and are blown away. When they are deposited on the human skin or body of the animals, they may create allergic reaction.

Flowering starts in about a month's time after seed germination and fresh growth comes up and grows round the year under favorable moisture conditions. Initiation of inflorescence (head or capitulum) is terminal, scorpiod and seems to terminate the vegetative growth of the plant. Inflorescences borne on branched peduncles held at the top of the plant are hairy and give a whitish appearance. Florets are creamy white, outer florets 5, female supported by two hyaline wings at base; inner few male florets, corolla in female ligulate; in male tubular,

5-lobed and anthers obtuse below. Flowering and fruiting continue profusely throughout the year and the plant is also capable of withstanding heat wave and mild frost. In nutshell, the plant has the capability to grow in all types of climate and soils. Parthenium is a C_3–C_4 intermediate plant and has low photorespiration, which makes it an efficient plant. Pollen grains are relatively small, seen in clusters and lighter than those of other species of Asteraceae which have been included in the group of allergens and anemophilous, the five peripheral ray florets produce the cypselae (fruits). Fruit is dorsally compressed, narrow below with pappus of two lateral refluxed awns.

The pollen grains are typically 3-colporate, oblate-sphereoidal, spinulose, a characteristic feature of majority members of Helianthae (sunflower tribe) and Ambrosieae (Agashe and Prathibha Vinay, 1975). A single plant of *Parthenium* may produce about 300 million pollens per square metre. The pollens are light in weight and are easily carried away by wind and water to a long distance. Parthenium has the potential of producing as high as 154,000 seeds/m^2 and a single plant can produce about 15–25,000 seeds (National Research Centre for Weeds-Annual Reports).

The collection of seed is rather difficult because of their small size and shedding during ripening. The seed coat imposed dormancy is broken in the natural course of weathering. The seeds remain viable on the soil surface for up to two years or longer if there is no rain to stimulate germination (Sukhada, 1975). Plants attain senescence after completing their reproductive cycle and skeletons of the dry and dead plants can be seen for a few weeks. Seeds readily germinate even with a little moisture availability and seedling regeneration occurs all round the year irrespective of the season. A detailed description of the species observed in India has been provided by Rao (1956). Phonological parameters of parthenium as observed by Tiwari and Bisen (1982) are given in **Table 1.** Further,

Table 1. Phenological parameters of *Parthenium hysterophorus* L.

Character	Days
Seedling emergence	7
Leaves attaining full size	26
Elongation of shoot ceases	70
Floral bud initiation	29
Flowering starts	31
Fruiting starts	36
Fruit maturity starts	57
Flowering ends	78
Leaf senescence (lower leaves)	37
Leaves begin to fall	41
All the leaves have fallen	103
Complete plant maturity	123

Source: Tiwari and Bisen (1982)

Table 2. Morphological and reproductive characters of Parthenium

	*	**
Plant height (cm)	101.5	50-150
Stem diameter (cm)	1.8	1.5-2.3
Branches/plant (main)	5.2	4-11
Leaf number/plant	38.6	21-43
Root length (cm)	20.0	13-24
Number of heads/plant	184.6	79-371
Number of seeds/head	5.0	—
Number of seeds/plant	923.0	700-4000
1000 seed weight (mg)	400.0	350-450
Seed germination (%)	75.8	15-81

Source: *Tiwari and Bisen (1982); **Anon. (1991)

its morphological and reproductive characters are furnished in **Table 2.**

Seed longevity: Long term burial studies have shown that 74% of the Parthenium weed seed were viable even after 2 years. The remainder either could not be recovered (18%) or did not germinate (8%). There was a log-linear decline in germinability of the buried seed over time indicating a constant rate of seed loss, with a half life of about 6 years (Navie *et al.,* 1998). This indicates that buried parthenium weed seed can remain viable much longer than previously thought (Butler, 1984).

Parthenium Biology and Distribution

Biology

P. hysterophorus is native to the countries bordering the Gulf of Mexico, and has spread throughout the southern USA Mexico, the Caribbean and Brazil (Navie *et al.,* 1996) (Biological Control of Parthenium through *Zygogramma biocolorata*, NRCWS Technical Bulletin No. 5). A slightly different race of *P. hysterophorus*, with yellowish flowers, is native to central South America (Argentina, Bolivia, Chile, Peraguay, Peru and Uruguay). The probable geographic centre of origin is the countries around the Gulf of Mexico.

Present distribution

In India, *P. hysterophorus* occurs throughout the country except in west coast of southern peninsula, from Kanya Kumari to Goa. The central and north Indian states are the worst affected ones in the country (see map). These are involved in active eradication campaign, but in Mussorie, the infestations are so numerous that the situation is becoming critical. Parthenium grows in all States of Australia. In the Northern Territory, this weed

Source of information about Parthenium

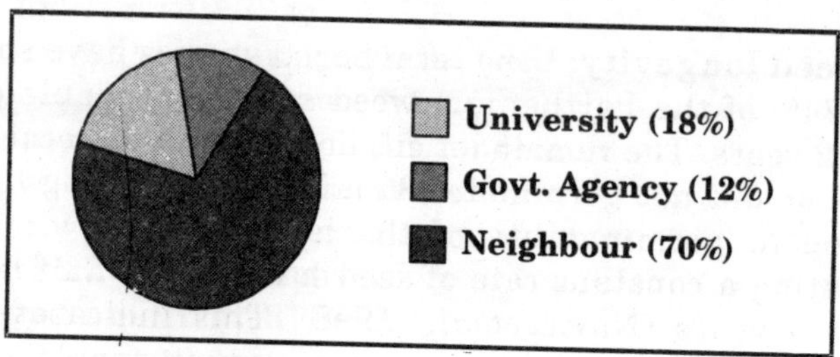

Modes of introduction of Parthenium in the area

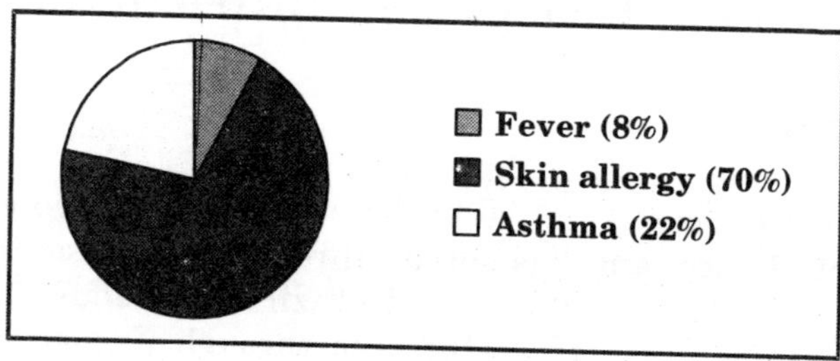

occurs along a stretch of the Roper Rover but has not spread significantly further from this. The plant grows on a wide range of soil types from sandy to heavy clays, but favours the latter. It is likely to grow in any area with a summer rainfall greater than 500 mm/annum.

Parthenium also occurs in Taiwan, southern China, the Pacific Islands, and India (Navie *et al.*, 1996) and has recently spread to East and South Africa.

Chemical composition

Since *Parthenium hysterophorus* causes ill-effects on man, livestock and other plant species, it is important to know its chemical constituents and in this regard a few studies are reviewed here.

Parthenin: Sesquiterpene lactones derived from plants of the family *Asteraceae* have been reported to be responsible for contact dermatitis (Mitchell and Dupuis, 1971). By extracting the plant with dilute alcohol, Arny (1890, 1897) was able to isolate an apparently non-alkaloid, non-glycosidic substance **"Parthenin"**. The yield of this crystalline material **(Table 3)** which accumulates in the leaves appears to vary seasonally and reach a peak in late summer (Herz *et al.*, 1962). In an earlier study, Herz *et al.* (1959) obtained 0.14 % parthenin from the entire plant. In a subsequent study, the yield of parthenin was 0.33 % of the plant dry weight. Parthenin is a sesquiterpene lactone psuedoguianolide (Kelkar, 1970) with abnormal carbon skeleton (Herz *et al.*, 1962). This active chemical is known to cause dermatitis (Ranade, 1970). Besides, it also contains water soluble inhibitors like caffeic acid and coumaric acid (Lakshmi Rajan, 1973) as well as anesic, vanillic and hydroxyl benzoic acid (Kanchan, 1975). All these are present in varying concentrations in different plant parts, being high in leaves followed by inflorescence, fruits, root and stem (Kanchan, 1975). Some more points on these aspects are also covered elsewhere in this book. While Parthenin is the major sesquiterpene lactone present in the plants growing wild in India, in some races growing in Mexico, parthenin is replaced by its diasteriomer hymenon.

Table 3. Distribution of parthenin in *P. hysterophorus* as determined by high performance liquid chromatography

Plant part	Amount of parthenin (% wet weight)
Leaf	3.40
Flower	1.08
Stem	0.12
Root	0.00
Trichomes	1.20

Source: Subba Rao *et al.* (1976)

Allelopathy and its effects on plant species

Allelopathy is the inhibition of growth of a plant due to biomolecules (allelochemicals) released by another. The biomolecules are produced by some plants as secondary metabolites. When the allelochemicals are released into the environment, they inhibit the development of neighbouring plants. Although allelopathic science is a relatively new field of study, there exists convincing evidence that allelopathic interactions between plants play a crucial role in both natural and manipulated ecosystems. Allelopathic interactions are also thought to be an important factor in the success of many invasive plants.

Apart from its competitive ability, the invasiveness of the noxious weed *Parthenium hysterophorus* is thought to be due to its ability to displace other species by means of allelopathy. The sesquiterpene lactones parthenin that is biosynthesized by this species is thought to play a role in its allelopathic interference with surrounding plants. However, despite the fact that parthenin is released from various plant parts into the soil, a little is known about its relative contribution to overall allelopathic effects. As leaf residues are believed to deliver large amounts of parthenin to soils during decomposition, the level of involvement of parthenin in overall phytotoxicity of decomposing leaf material was investigated in a South African population of *P. hysterophorus*. Furthermore, parthenin treatments proved to significantly delay germination and stimulate root growth at low doses. The contribution of parthenin was highly dependent on its concentration within extract solutions and varied between 16% and 100% of overall phytotoxicity of leaf extracts.

This suggested that the release of parthenin during decomposition of leaf material has a potential to play a leading role for allelopathy in *P. hysterophorus*; however, its significance in a natural setting very much rely on the amount of leaf material accumulated on soil surfaces and the concentration of parthenin in residues.

Bioassay, pot culture, and field studies have revealed that all plant parts namely shoot, root, inflorescence and seed are toxic to crop plants.

Growth pattern

The growth of Parthenium was observed in and around Bangalore city (India) throughout the year where the commencement of rains differed in different parts of the area. Parthenium plants could establish and attain maturity with just two rains in the month of February as observed during 1988 and 1990. It could also establish with the help of a few showers received in the month of May in all the four years, in pockets where rains were received. Since the showers during the month of May start in different weeks in different years, it provided a good opportunity for recording the growth pattern of the plant. Irrespective of the month of commencement of the rain, Parthenium seeds germinated and the plants attained maturity in 130 to 165 days. Water used for road repairs, continuous washing of cattle in the yard and open toilet places is seen with parthenium growth round the year in villages and city limits. Thus, Parthenium colonies could be seen in all the stages of its growth throughout the year.

Growth of Parthenium was higher in cropped lands than in non-cropped lands (Tiwari and Bisen, 1982), while under irrigated cropped lands, *rabi* population is more vigorous. However, in places under non-cropped land due to winter rains in December the growth is poor and the survival rate is 30.5% and most of them perished at the seedling stage due to moisture stress. The plants hibernate in “rosette” form till adequate rains occur. Upward curling of leaves is common during summer but after the rains, such plants resume growth and flower normally.

Parthenium in eco-systems

After noticeable occurrence of this plant (*Parthenium hysterophorus*) in 1955 in Pune (Maharashtra) in India, it was not considered a serious weed till 1975. It was Gidwani (1975) who brought the attention of Indian scientists towards this problem with his article that appeared in the Newspaper "Current". He estimated Parthenium invasion in the country at that time to be in 5 million hectares of land.

Parthenium grows wild only is some countries in a variety of ecosystems with luxuriant growth around the year. High altitudes like Kollimalai, Ooty and Annamalai hills in Tamil Nadu, Bhagamandala, Biligiri Rangana Tittu and Baba Budangiri Hills in Karnataka and Kulu valley in Himachal Pradesh harbor this weed. It does not show wild growth in hilly regions and in high rainfall areas like south Kanara and North Kanara in the state of Karnataka and along the coast in Kerala and Orissa states. However, it is found growing in the east coast of Tamil Nadu and Andhra Pradesh. It is also not a serious problem along the coastal line of Goa. Attempts in Karnataka state to grow this near Mangalore and Udupi for research purposes were not successful (Mahadevappa *et al.,* 1990). Under the conditions of very high rainfall, the seeds do not set properly. It is absent in the hilly areas of Maharashtra and Tamil Nadu and also in the laterite soil areas around the city of Bhubhaneshwar in Orissa. Whether the factors like lateritic soil condition and low temperature inhibit germination and growth of this weed needs to be verified. Parthenium can adapt to varying cropping and ecological situations. It also appears to have its own adaptive protective measures to tide over water stress (Tiwari and Bisen, 1982).

The Parthenium is all pervading in majority of the agricultural lands (Plate 2). In paddy fields, it is seen only on the bunds but not in the puddled submerged area.

It grows along with the crop in drilled paddy fields. Even in rocky areas, even with a little soil deposited in the rock crevices and joints, it can put up normal growth. Krishna Murthy *et al.* (1977) reported its growth even in the land masses isolated by water as in the case of Ulsoor tank in Bangalore city. Limited growth was observed by the author in hilly areas like Malai Mahadeshwara and Biligiri Rangana hills in Karnataka state, India. In a biological survey conducted in late eighties (Mamatha and Mahadevappa, 1988), intensity of Parthenium has been observed to be very high with bushy growth in Uttar Pradesh, Karnataka and Maharashtra. Parthenium growth is dense in and around cities, townships and villages (Plates 3 & 4) and its intensity gradually declines as we proceed to areas not inhabited by human population. This means "more the human activity more the Parthenium invasion".

Parthenium population is high in places where the soils are being disturbed constantly for purposes of construction of buildings (Plate 5), roads and lifting of soils, etc. In cultivated fallows (Plate 6) of agricultural lands, it is seen abundantly but not in the regularly cropped fields. In orchards (Plate 7), where ploughing is done once in a while, the weed grows luxuriantly. Maheshwari, (1966) has stated that this has become a neo-tropical weed which is highly undesirable and can cause much damage.

The floristic studies of *P. hysterophorus* communities under non-cropped and cropped lands were conducted during *kharif* 1980 and *rabi* 1980–81 by Tiwari and Bisen (1982). In non-cropped lands *P. hysterophorus* had the highest stand in *kharif* followed by *Cassia tora*, *Acanthospermum hispidum* and *Echinochloa crusgalli*. The floristic composition of *Parthenium – Argemone – Dichanthium – Malvastrum* type community was found developing following germination of Parthenium in winter. The stand of Parthenium was greater in *rabi* than *kharif*.

Under cropped lands during *kharif*, Parthenium was noted in the fields of soybean, millets and paddy indicating its adaptability to different soils and moisture levels. A community of *Echinochloa – Digitaria – Parthenium – Acanthospermum* type was found under these habitats.

Parthenium in Plant Community

Whether parthenium will stay on for ever if no effective measures are developed to suppress it, is a question that may arise in the discussions like this. Any single species dominating a specific habitat has always had a successor. It may be mentioned here that there are quite a few instances reported wherein the plant succession is caused

Table 4. Interference among waste land species and their impact on parthenium suppression

Species	**Extent of parthenium suppression***
Amaranthus spinosus L.	High
Croton bonplandiamum Baill	High
Cassia occidentalis L.	High
Cassia tora L.	High
Cassia sericea SW.	Very High
Cassia auriculata L.	Moderate
Hyptis suaveolens Poit.	Moderate
Ipomoea carnea Jacq.	Very high
Mirabilis jalapa L.	Very high
Sida spinosa L.	Moderate
Tephrosia purpurea Pers.	Very high

*Moderate: 26–50%, High: 51–75%, Very high: 76–100%
Note: All the above spices are known to suppress Parthenium through allelochemicals, except *Ipomoea carnea* which is a physical suppresser.

because of the allelopathic impact of succeeding plant community on the preceding one (Rice, 1984), the same way as *cassia sericea* is replacing Parthenium in waste lands. The pattern of invasion, extent of establishment and interference of parthenium in the waste lands is presented in **Table 4.** But parthenium has several built-in properties and efficient mechanisms which enable it to overcome many ecological adversities and thus continue to survive defying law of natural succession of wild flora. Hence, in the absence of any effort, the parthenium plant may be able to survive comparatively a longer period than did its predecessors which dominated in the past. This is not desirable. The entry of *C. sericea* in the natural course itself is a clear example that, as, in the history, nature's action has already begun to replace parthenium also. But it is going to be a rather very slow process. It may now be recalled that promoting the growth of *C. sericea* is actually going in the nature's way and in other words definitely not going against the natural course and hence will be ecologically sustainable and environmentally safe. Promoting the growth of *Cassia sericea* and similar bioagents is only promoting the nature's mechanism to allow natural succession. Logically, it will be eco-friendly and sustainable.

Parthenium in Agricultural Waste Lands

Omnipresence of parthenium in agricultural and waste lands is discussed in detail elsewhere. Depending on the fertility and the rainfall, its growth will be either stunted or moderate or vigorous but at least a stunted growth of this weed can be seen when one or two rains are reconceived during the fallow periods. In orchards of guava, coconut, grapes, sapota, mango, cashew, papaya etc., this weed grows profusely because, in such plantations, weeding is not done as frequently and systematically as is done in the continuously cropped fields, thus, doubling cultivation costs besides reducing crop yields **(Table 5)**. The occurrence of this weed in grasslands is reported to reduce forage production up to 90 % besides making the land less fertile (Vartak, 1968). The weed has also been seen in association with groundnut, potato and cotton and is reported to have hampered their production considerably in Karnataka (Hosmani and Prabhakar Shetty, 1973). The authors' impression is that the reported losses due to this weed in agricultural field are over estimates.

Some studies have also been made on the effect of parthenium plant parts on the agri-horticultural crops. The dried and powdered plant materials when incorporated into the soil have been reported to inhibit the germination, growth and yield of wheat (Lakshmi Rajan, 1973) and ragi (Kanchan, 1975). Besides, the water extract from the roots reduced the growth and colonization of nodule bacteria. Further, the yield of tomato and fingermillet (*Eleusine coracana*) was reduced by 40 to 50% due to decrease in the number of branches and tillers, respectively. The seed germination of cowpea and beans was reduced to the extent of 50 and 60%, respectively, over control in the presence of extracts prepared out of the inflorescence and the leaf of the weed. However, in the presence of the extract of stem and the roots, the reduction in coleoptiles height and dry weight was considerably less (20 to 25% and 40 to 45%, respectively over control). Though the coleoptiles height of wheat was not affected much by the extract of fruits (the reduction, being only 25 %), yet its effect on plant dry weight was severe (reduction being 75 % over control) (Lakshmi Rajan, 1973). Some information relevant to this chapter is already covered under "Parthenium in ecosystems". Jeevan Rao and Shantaram (1997) conducted a survey around Hyderabad in seven farmers' fields for Parthenium infestation where urban solid wastes were used as manure for more than a decade. The fields were highly infested with the weed. The growth of crops like tomato, lady's finger (bendi), egg plant (brinjal), bottle guard, radish

Table 5. Effect of parthenium on growth and flowering of field (50DAS) and crop yield loss due to uncontrolled parthenium

Field crops	**Parthenium population (m^{-2})**	**Parthenium weight (g)**		**Parthenium flower heads**		**Crop yield loss**	
		(m^{-2})	**%**	**No.**	**%**	**kg/ha**	**% loss**
Sorghum	11.31 (1.05)	127	87.3	1256	61.2	445	14.7
Pearlmillet	15.0 (1.180)	195	80.5	1410	56.4	394	23.3
Maize	10.3 (1.01)	127	87.3	1058	67.3	416	12.3
Sesamum	21.0 (1.32)	315	68.5	2043	36.9	75	25.0
Groundnut	25.0 (1.40)	425	57.4	2216	31.5	342	27.4
Sunflower	12.3 (1.09)	155	84.5	1383	57.2	157	14.1
Urdbean	24.0 (1.38)	416	58.3	2402	25.8	192	23.8
Mungbean	21.3 (1.33)	300	70.0	2531	21.8	187	21.2
Soybean	18.6 (1.27)	261	73.8	2466	23.8	231	21.3
Cowpea	22.3 (1.35)	364	63.7	2382	26.4	24	20.1
Pigeonpea	26.0 (1.41)	494	50.6	2690	16.9	206	20.4
Control (Parthenium only)	28.0 (1.45)	999	—	32.37	—	—	—
CD (P=0.05)	0.16	160	—	106	—	—	—

Figures in the parentheses are log transformed values. *Source*: Kandasamy and Sankaran (1997)

and turnip were largely affected by Parthenium infestation. Parthenium samples were analyzed for heavy metals. The shoots contained more heavy metals than roots **(Table 6).**

Table 6. Mean heavy metal content (ppm) of parthenium samples collected from soils treated with urban solid wastes

Heavy metal	Root	Shoot
Iron	365.00	665.00
Zinc	259.00	85.00
Copper	10.00	11.00
Manganese	19.00	73.00
Lead	11.00	60.00
Nickel	1.80	4.60
Cobalt	5.50	0.55
Chromium	0.45	2.00
Cadmium	2.50	2.00

Source: Jeevan Rao and Shantaram (1997)

Three

USE AND MISUSE OF PARTHENIUM

In the genus *Parthenium*, only one species, *Parthenium argentatum* Gray (guayule), is of potential economic value. Guayule has frequently been suggested as a potential source of natural rubber, but has never been commercially exploited.

Parthenium hysterophorus has been regarded as a beneficial plant too, both in India and abroad, in spite of its ill-effects which outweigh its uses. The word Parthenium is derived from the Latin word 'parthenice', suggesting medicinal uses. Its medicinal use has been reported as early as 1897 by Arny in the West Indies. It is used as a folk remedy against various afflictions such as ulcerated sores, certain skin diseases, facial neuralgia, fever and anemia (Arny, 1890, 1897). According to Curtis (1921), dry flowers are used as tonic for digestion, cleaning the blood, abortive, vermifuge, emenagogue and as insecticide in parts of Europe. Positive tests of extract of Parthenium plant with Wagner and Dithmar reagent is reported. They indicated that the medicinal properties, if any, must be due to substances other than alkaloids. They observed very small amount of alkaloids from the Parthenium plant. Uphof (1959) has reported on its root decoction being used by Kosti Indians as a curative for dysentery and some kinds of fever. There are also opinions expressed in India on its insecticidal value and for curing skin diseases like psoriasis. According to Kannabiram (1975), studies made at the Cancer Research Institute, Mumbai, parthenin, the principal ingredient, possesses anticancer properties. When such a property is confirmed, its profitable exploitation by pharmaceutical industry to produce the anti-cancer drug could be explored. All such vague suggestions have been criticized both in and outside India. As felt at present, its harmful effects both from medical as well as animal health and agricultural production point of view are serious matters for consideration (Krishna Murthy *et al.*, 1977).

Parthenium hysterophorus showed significant reduction in blood glucose level in the diabetic ($p < 0.01$) rats. However, the reduction in blood glucose level with aqueous extract was less than with the standard drug glibenclamide. The extract showed less hypoglycemic effect in fasted normal rats, ($p < 0.05$). It was concluded that the active fraction of *Parthenium hysterophorus* flower extract is very promising for developing standardized phytomedicine for diabetes mellitus (Patel *et al.,* 2008).

Of late, parthenium plants have been commonly used as centering material in the construction of buildings and bridges in several parts of Karnataka and neighbouring states. Parthenium flowers/plants are also used for decorations on festive occasions either singly or mixed with other flowers and plants (Plate 33a). Some people out of ignorance or otherwise are using parthenium flowers for preparations of bouquets (Plate 33b) in flower vases etc. Further, studies carried out at the Indian Institute of Science, Bangalore during mid-eighties have indicated a possibility that Parthenium plants can be ensilaged after mixing it in certain proportion with conventionally used feeds (Narasimhan, 1987, personal communication). The fibrous nature of the parthenium stem is a point that needs the attention of industrial researchers to see if it has potential in manufacturing paper, chip board or such other materials. In Europe, the sister species of Parthenium *viz.*, *P. argentatum* is reported to have potential economic value and is being used in the manufacturing of tyres for aero planes. The weed is also used as firewood, as green manure in rice fields and as packing material for guava, sapota, apple and tomato. All these, apart from being hazardous, add to the spread of the weed. The leaves of Parthenium which have close resemblance to those of "Davana" (*Artemisia pallens* Wall) are mixed with it and sold. When ladies wear, it causes irritation on neck and cheek (Personal communication by Dr. Bir Bahadur, professor of Botany, Warangal). Parthenium leaves are also seen bundled with coriander leaves and sold in the vegetable market. Thus, it is being used for adulteration in market places. In Khajuria Kala, a village in Ashoknagar district, a newly-created district of Madhya Pradesh, India which houses one of the biggest Mandis in the state, this plant is used for preparing fertilizers, for a long time (IRPNG Website).

It has also been reported that the solid matter of the parthenium being fibrous, it is possible to convert it into natural fiber reinforced polymer (NFRP) composites and these parthenium based NFRP composites can be manufactured and converted to end-products like helmets, moulded tableware, moulded chairs, moulded flooring, moulded doors etc.

Pollen grains of Parthenium also possess antifungal activity by producing an inhibitory effect ion downy mildew, *Sclerospora graminicola* (Sacc.) Schroet (Char and Bhat, 1975a). Here, zoospore, motility and sporangial germination were inhibited. The rate of inhibition of sporangial germination by pollen was highly significant. Though the sporangial germination and zoospore motility in *S. graminicola* was inhibited due to parthenin, it did not exercise the same activity in greater concentrations. Thus, it may be inhibitory to one organism but may not exercise the same activity on another (Char and Bhat,

1975b). However, slight antibacterial and absence of antileukemic activity of parthenin has already been noticed.

Patil *et al.* (1977a) reported that 20% leaf water extract of Parthenium, when sprinkled on mulberry leaves before collecting for feeding, improved both the qualitative and quantitative traits like larval weight, fineness of silkworm thread, cocoon weight, (Plate 8) etc. and attributed the improvement to the similarity of steroids present in the Parthenium plant. Hiremath and Ahn (1997) reported that Parthenium extract in methanol had insecticidal properties on brown plant hopper of paddy. Nagaraja *et al.* (1997) observed stimulating effect of Parthenium leaf extract on soybean and chickpea. The effect of control of eupatorium weed by Parthenium leachates application was reported by Patil *et al.* (1977b) and Chetti *et al.* (1997). In sorghum, Parthenium extract spray was found to reduce the shoot fly incidence by 17% (Bhuti Hiremath, 1997). In redgram, the pod and grain damage was reduced by treatment with Parthenium (Da *et al.*, 1977). Parthenium along with cow dung could be used in the production of bio-gas although the yield was lower than when water hyacinth plants (Sreenivasa and Majjigudda, 1997). Thimmaiah and Bhatangar (1997) developed a modified substrata involving chopped Parthenium leaves (50g or 100g) and cow dung manure (300g) for successful vermitechnology. Son and Ramaswami (1997) experimented on the decomposition pattern and nutrient enhancement of Parthenium and found that composted Parthenium recorded higher phosphorus content while composted sugarcane trash recorded higher potassium content. The highest percentage of reduction (41.45%) on phenol content was higher in parthenium even though the phenol content was higher in Parthenium at the end of composting. The weed population in the rice field was influenced by the incorporation of organic wastes like composted Parthenium, coir pith and Parthenium. In residual crop of soybean, composted parthenium application recorded higher grain yield over inorganic and other organic wastes. Sreenivasa and Majjigudda (1997) studied the possibility of utilizing common weeds including parthenium for cassia and water hyacinth in various combinations with cattle dung for biogas generation to find that cassia and Parthenium in combination with cattle dung at 25:75 (w/w) resulted in 156 and 137 litres of biogas production per kg of dry matter, respectively. Higher proportion of parthenium in the mixture was found to be deleterious which resulted in reduction in biogas formation as well as methane content.

Kohi *et al.* (1997) conducted a study that the plant extract of Parthenium to control the weed, *Ageratum conyzoides* from assessing its effect on seed germination, seed vigour, water content, chlorophyll content, cellular respiration, biomass, leaf temperature, relative humidity of leaves and transpiration rate and found that germination and seed vigour decreased linearly with the increase in treatment concentration compared to control. Photosynthetic efficiency, pigment content as well as the cellular respiration was also reduced. However, biomass remained little affected as compared to control. Leaf temperature was lowered with the parthenium spray. Relative humidity was slightly high in the lower concentrations but was similar to control in higher concentrations. Transpiration rate was high in the treated plants. Thus, parthenin adversely affects the metabolic make-up of *A. conyzoides* and can be tried to check its seed germination and growth. The sesquiterpene lactones parthenin that is biosynthesized by this species in reported to play a role in its allelopathic effects. Similar study made by Chetti *et al.* (1997) using parthenium extract to control eupatorium revealed significant differences on over-all biophysical and biochemical constituents due to Parthenium plant extracts and concentrations. They significantly reduced relative water content, photosynthetic rate, transpiration rate, chlorophyll content, and sugar content and phenol contents. Leaf extracts of Parthenium were more effective in reducing the above parameters than the root extracts. Increasing the concentration of extract from 5 to 10% significantly decreased the biophysical and biochemical parameters.

Detoxification

It is reported that Parthenium could be detoxified by insulation *i.e.*, anaerobic fermentation. This was done by packing chaff-cut plants in trenches along with small quantities of salt and allowing fermenting for 2 to 3 months. The silo thus obtained was reported to be devoid of the contact allergen/toxin parthenin and was highly palatable to live stock. The animals did not show any ill-effects even after feeding with the silo for 3 months. The results of nutritional evaluation summarized recently revealed that the Parthenium silo is comparable to the well known fodder silos with high nutritive value (Narasimhan *et al.,* 1977). This however, needs verification through large scale experiments on various animals of different ages and using the parthenium plants grown in different agro-climatic situations.

Four

SPREAD OF PARTHENIUM

In India

In India, it was first observed in Poona in 1955 as a stray plant on the rubbish heaps in the neighborhood of the Agricultural College, Poona by Prof. H.P. Paranjape, Retired Horticulturist of the then Bombay State (Vartak, 1968). This was identified and described by Rao (1956) as *Parthenium hysterophorus* L. Subsequently, the identity of the Parthenium and its occurrence as a new adventives plant in Poona was confirmed by several botanists (Lonkar *et al.*, 1974). The same species was also found growing abundantly near the river bank and scattered all along the dried fields in and around Poona as well as in the agricultural fields (Rao, 1956). Even then its bad effects were not known, but, in a couple of years, it has spread very fast all over Poona covering waste lands, railway yards, marshy patches, unused cultivable lands, grasslands, roadsides, along the canals and other areas (Maheshwari, 1966; Santapau, 1967). The 1958–59 floods in Poona are believed to have enabled the spread of this plant initially (Roberts, 1967). Subsequently, after Panshet dam, in the floods that broke out during 1961, most of the grain go-downs near the river banks were washed out. Linked with this was the rapid spread of this plant all over Maharashtra in a matter of few years (Ranade, 1975). By 1965, it covered most of the agricultural lands in Maharashtra (Vartak, 1968). Evidently, it spread to other parts of the country from Maharashtra through vehicular traffic, trains, cargo, packaging material, besides other natural agencies of dispersal.

The weed, noticed first in Maharashtra, gradually spread to Karnataka. Its first appearance in Karnataka was near Dharwad (Ladwa and Patil, 1961). Maheshwari (1966) observed parthenium as a stray plant growing near the outskirts of Delhi at Moti Bagh and along with Delhi ridge in 1964. The weed showed a very high numerical increase in the construction plots of these settlements by October, 1965 (Maheshwari, 1966a). By 1969, it became a completely naturalized weed in Delhi (Vaid and Naithani, 1970). By mid-sixties, the weed is reported to have entered Kashmir possibly from Poona along with some Jasmine root cuttings (Hakoo, 1963). Subsequently, the weed made further inroads into central parts of the country and was spotted in Itarsi in Madhya Pradesh during 1968. It is believed to have reached through timber piles obtained from Dandeli in Karnataka (Maheshwari, 1968). It spread in alarming proportion in Tamil Nadu in late sixties, in Andhra Pradesh during early 1970s and in Uttar Pradesh during 1980s. The author observed the most luxurious growth of this plant in Haridwar (Uttaranchal) during late eighties.

It is estimated that about 5 million ha of land has been invaded by parthenium in several states and the union territories in the country. The spread of Parthenium in India is given in **Table 7.** A statewide picture of Parthenium spread as observed or reported by various authors is presented below.

No reports are available about the status of Parthenium in Andaman and Nikobar islands.

In **Andhra Pradesh**, parthenium was reported to be present in Hyderabad, Krishna, Godavari and Chittor

Table 7. Spread of Parthenium in India

State	Areas	References
Andhra Pradesh	- Hyderabad, Krishna, Godavari, Chittor and Nagarjun Sagar areas.	Krishna Murthy *et al.* (1977)
	- Tirupati hills and Khadala	Santapau (1967)
	- Amaravati, Nalla Malais, Karnool, Mahaboonagar, Thimmapur, Waltair, Vijayawada and Moosa river	
Bihar	- Mothihari, Narkatiarganj, Balmikinagar	Maheshwari and Pandey (1973)
	- Gangetic plains of Bihar	Suresh Chandra (1973)
Gujarat	- Ahmedabad, Anand and Baroda	Mahadevappa (1996)
Haryana	- Rohtak, Hissar and Faridabad Eastern Parts of Haryana	Mahadevappa *et al.* (1990)
Himachal Pradesh	- Kulu and Manali	Vaid and Naithani (1970)
Jammu & Kashmir	- Throughout the state (Personal communication)	Prabhakar (1988)
Karnataka	- Dharwad	Mahadevappa *et al.* (1990)

Table 7. *Continued ....*

State	Areas	References
	- Banglalore, Mysore, arsikere, Belur, Bhadravati and Shimoga	Jayachandran (1971)
	- Kodagu, South Canara and North Canara	Mahadevappa (1996)
	- Bijapur, Belgaum, Gulbarga, Bidar,	News' paper publications
	- Chitradurga, Shimoga, Hassan, Mandya and Tumkur	
Kerala	- Palghat, Quilon, Kottayam and Kasargod	Mahadevappa (1996)
Madhya Pradesh	- All over the state except in hills	Tiwari and Bisen (1984)
Maharashtra	- Mumbai city (juhu area)	Tiwari and Bisen (1984)
	- Forest nurseries	Chandras (1970)
	- Joshi estate	Mahadevappa (1996)
Orissa	- Not much, only in the peripheral parts	Mahadevappa (1996)
Punjab	- Many parts of Punjab	Mahadevappa (1996)

Table 7. *Continued ....*

State	Areas	References
Pondichery	- Scattered distribution throughout the state	Mahadevappa (1996)
Rajasthan	- Udaipur	Mahadevappa (1996)
Tamil Nadu	- Kotagiri of Nilgiri hills	Bidhas Ray (1975)
	- Aliyar submergenic area, Coimbatore Katpadi, Jolarpet, Madurai, Salem, Tanjore	Mahadevappa (1996)
Uttaranchal	- Pantanagar, Rai Bareli, Jhansi	Elis and Swaminathan (1969)
	- Haridwar	Mahadevappa (1996)
West Bengal	- Kolkata, Bank and Basin of rivers	Krishna Murthy *et al.* (1977)

districts. There is also a report on the occurrence of Parthenium in abundance in Nagarjuna Sagar areas (Krishna Murthy *et al.*, 1977). Further, a few plants were also observed on top of Tirupati hills (author's observation, 1987). The weed was also observed on Khadala (Santapau, 1967); Amaravati, Nalla Malais, Karnool district, Mehaboobnagar, Thimmapur, Waltair, Vijayawada and Moosa river in Hyderabad.

In **Bihar,** Maheshwari and Pandey (1973) observed the occurrence of Parthenium in Mothihari, Narkatiagani town and Balmikinagar in Champaran district of Bihar. Suresh Chandra (1973) indicated the entrance of the weed in the alluvial soils of the Gangetic plain and found it growing abundantly in various parts of the state.

In **Gujarat, t**his weed was observed by the author in Ahmedabad during 1972 on road sides along the highway between Ahmedabad and Anand. It was also seen growing in greater proportion in public and agricultural lands around Baroda.

In **Goa,** the author made three visits along the coastal line and could not see Parthenium.

In **Haryana,** this plant was found to be more towards areas bordering Delhi than in the northern areas as noticed by the author in 1990. Its growth was luxurious towards Uttar Pradesh as compared to that in the northern parts of the state.

In **Himachal Pradesh, t**his plant was found spreading deep into hilly areas and was found in abundance near Manali (Kulu valley) (Vaid and Naithani, 1970). At present, this obnoxious plant has covered whole of Una, Hamirpur, Bilaspur, Mandi, low and mid-hill zones of Chamba, Kangra, Solan, Srimaur, Kulu and Shimla districts of the state (Angiras and Saini, 1997). Intensity of this plant has been observed to be more around areas of hydro-electric projects, road sides, wastelands, forest lands, pastures and grasslands. Dissemination has been observed to be more through interstate transport machinery and wind. Observations have also indicated that invasion of this plant is less in areas occupied by ***Cassia tora*** in the natural system. Parthenium has invaded majority of apple orchards in lower elevations (Bisht, 2004). The invasion of Parthenium was reported in forest and wastelands with little or no growth of any other species and local bio-diversity was found to be threatened (Kumar and Rohatgi, 1999). In many forests, National Parks and Reserved forests, the occurrence of this weed has been noticed.

In **Jammu and Kashmir,** a state of hill resort is also on India's map of parthenium infestation. This weed is said to have entered J & K some-where in 1963 from Madhopur in Punjab, across river 'Ravi', all along the national highway. In contrast, this plant made fast entry with the extension of railway track to Jammu from Punjab. Subsequently colonization occurred throughout vacant places and then it spread in all directions along man-disturbed habitats covering many hundred kilometers of road distance up to Poonch and Kashmir, the north-western border adjoining Pakistan. Current situation in the state is quite alarming as the plant has engulfed vast areas into the interior of forests and all vacant lands in the vicinity and outskirts of inhabited places.

In **Jharkhand** cases of breathlessness and asthma due to pollution through different sources *viz* air, water and pollen may rise alarmingly in the city of Ranchi the capital of Jharkhand State of India. Threat comes from rapid growth of Parthenium grass, better known as "Congress Grass". People of Ranchi are affected with asthma, allergic, trinities' sinusitis, dermatitis (type of skin disease) especially among the children, eczema, allergic papules and all types of allergic reactions. It cannot be said that above mentioned diseases were not common in Ranchi earlier. Climate of Ranchi unfortunately favors such types of diseases, but with the growth of Parthenium these diseases are affecting more people. Just few years ago Ranchi was free from such toxic plants but now this weed has shown its presence threatening the health of the Ranchi people.

In **Jharkhand**, the most vulnerable areas are outskirts of Ranchi city like HEC Township, SAIL and MECON and Harmu colonies. These are places where the weed is found in plenty and it has been seen spreading to other parts of the city. The most interesting is that this weed has also surrounded Jharkhand State Pollution Control Board office in HEC area. This plant has covered most parts of the HEC areas, especially along the roadsides, and the waste lands around J.N. College campus, and it is rapidly moving towards the center of the city. This spreading is going to cause many serious health problems to the people of Ranchi. If it intrudes into the agricultural domain, productivity is definitely going to be adversely affected. It squeezes grasslands and pastures, reducing the fodder supply. Scientists describe it as a "poisonous, allergic and aggressive weed posing a serious threat to human beings and livestock". The presence of Parthenium in cropped lands results in yield reduction up to 40%. It is also responsible for bitter milk disease in livestock fed on grass mixed with parthenium.

In **Karnataka**, though the weed was first noticed in Dharwad during 1961, Jayachandra (1971) reported the presence of its dense stands in Bangalore, Mysore, Arsikere, Birur, Bhadravati, Shimoga and many other stations on Bangalore – Talaguppa railway track. In a survey conducted by the author in Karnataka, the presence of Parthenium was observed in almost all districts in varying proportions. Patil *et al.* (1997c) conducted a survey of plants in Bijapur district and found that the infestation of Parthenium was more in irrigated as compared to rainfed areas. The infestation of parthenium was more in Mudhol, Jamakhandi and Bilgi tehsils in all the crops and was less in Sindagi, Bagalkot, Hungund and Bijapur district. It was further observed that the spread was more in sugarcane and plantation crops and least in vegetables. Parthenium has failed to establish in coastal Karnataka either in North Kanara or South Kanara districts although its growth in low intensity is seen in the borders of North Kanara adjoining Dharwad district. In hilly districts like Kodagu and Chikmagalur, its growth in thick population and in small colonies is seen in and around the townships and cities in constantly scrapped soils but not in the interior areas. Reports of its occurrence in the districts of Belgaum, Dharwad, Gulbarga, Bidar, Chitradurga, Shimoga, Hassan, Mandya, Mysore, Tumkur and Bangalore have been published in popular English and Kannada books, booklets as well as in news papers.

The intensity of parthenium growth is, however, very much reduced in the northern districts of Karnataka due to spread of botanical control agents like *Cassia sericea, C. tora, Tephrosia purpurea, Croton sparciflorus, Heptis suvaulensis, Sida spinosis* etc., and also due to release of *Z. bicolorata* (author's own observation).

In **Kerala,** this weed has recently spread in Palghat district. It is seen near railway stations in Quilon, Kottayam and within the cities though not in thick populations. It is also observed in Vakadi of Kasargod taluk of Kerala (author's observation). It has been noticed in Wynad region by the author in the year 2006.

In **Madhya Pradesh, t**he growth of Parthenium has been observed in menacing scale all over the state except in the hills (Tiwari and Bisen, 1984). The weed has also been observed at alarming rate in Pench National Park in Seoni District.

In **Maharashtra,** most parts of Mumbai metropolitan city are infested with Parthenium and the weed is particularly extensive in Juhu area (Tiwari and Bisen,

1984). It is also reported that the weed has become a menace in many forest nurseries in the state (Chandras, 1970). It was seen by the author in the pots along with medicinal plants in Joshi Estate, S.R. Road, Mumbai during 1990 and 1991.

In **Punjab,** Parthenium grows abundantly in many parts of Punjab especially in the rainy season. It is seen abundantly along the highways.

In **Pondicherry,** the growth of parthenium is much less in intensity as compared to the neighboring state of Tamil Nadu. But it is seen here and there round the year as observed by the author.

Parthenium is seen in Rajasthan also. It was noticed at Udaipur near the university campus in orchards and in some agricultural lands. It has become a big nuisance in Jaipur and Udaipur region besides its occurrence near the area irrigated by Indira Gandhi Canal throughout the year.

In **Tamil Nadu,** it has been reported to be present near Kotagiri (6600 ft above MSL) of Nilgiri hills (Baridhas Ray, 1975). The plant was observed in Aliyar submergence area and Coimbatore districts, including Katpadi and Jolarpet. The author has observed this weed all over Tamil Nadu but it is more visible in Madurai, Coimbatore and Salem districts and less in Tanjore and other districts adjoning Pondicherry.

In **Uttaranchal,** Parthenium, which used to be seen only on the road-sides in Asa Rodi forest of Dehàradun during 1995, has now made its way inside the forest also. The invasion of parthenium was at alarming rate during year 2000 in Rajaji National Park near Haridwar, Rishikesh and Dehradun (Goyal and Brahma, 2001).

In **Uttar Pradesh,** this plant was noticed at Pantnagar presently in Uttarkhand opposite to the railway station and has spread to a few agricultural lands, in Rai Bareli district and Jhansi areas (Ellis and Swaminathan,

1969). It grows most luxuriously in some districts especially around Haridwar presently in Uttarkhand as observed by the author. In Varanasi, Parthenium is in plenty in the nearby agricultural lands, on the bank and the basin of river Ganga (Krishna Murthy *et al.*, 1977).

In **West Bengal,** Parthenium is found in the very heart of Kolkata at the ladies golf club and cantonment areas. Further surveys conducted by Mukhopadhyay (1997) revealed that at present the highest density zones of Parthenium are (i) Salt Lake – Dhapa – Bantala – South Dum Dum areas near Kolkata city, (ii) Howrah – Dankuni railway tracks and yards, (iii) Durgapur – Asansol and neighboring coal field areas, (iv) Kalyani railway station and Food Corporation godown areas of Nadia district, (v) Farakka areas, (vi) New Jalpaiguri railway station track and yard areas in high density and luxurious growth was observed from July to September.

The record of Parthenium in Andaman and Nicobar Islands in Bay of Bengal is a pointer to the extraordinary ability of the plant to invade new environments.

With regard to its spread in other states in India, no documentary evidence could be gathered. At present, it is observed to be present in several states in alarming proportions attracting the attention of all concerned. Based on enquiries received by the author and author's own survey, it looks serious in Uttar Pradesh, Karnataka, Maharashtra, Tamil Nadu, Andhra Pradesh, West Bengal, Madhya Pradesh, Gujarat, Orissa, Bihar, etc. in that order. Further, it is not seen in the western coasts, high elevations and interior forests.

After Gidwani's (1975) estimate of Parthenium invasion in 5 million hectares, no effort has been made to revise the estimate of Parthenium infestation in India in spite of the fact that every researcher reported its severe occurrence in every type of land. As Parthenium is mainly a weed of fallow and wasteland, roadside, railway tracks, residential colonies and big industrial areas, the present infestation has been estimated keeping in view the total land use classification of India. The total geographical area of India is 328.73 million hectares out of which forests are 68.39 million hectares, land not available for cultivation (non agricultural uses + barren and uncultivable land other than fallow land is 29.09 million hectares, fallow land including current fallow accounts for 23.30 million hectares and gross cropped area is 188.15 million hectares (Anonymous, 1999). In general, fallow, barren, cultural wasteland, permanent pastures and other grazing land make up about 71.16 million hectares. Such lands are most suitable sites for the growth of parthenium abundantly ranging from 0 to 100% depending on the climatic conditions. In a very conservative estimate, if only 10% area is taken under Parthenium infestation, it alone comes to the tune of 7.11 million hectares. Area under agricultural and non-agricultural uses is about 188.15 and 22.51 million hectares, respectively. At present, every type of crop is infested with Parthenium and infestation is more severe in horticultural land. If we take only one percent of the crop land and non-agriculturally used land under Parthenium infestation, it comes to about 2.11 million hectares. There are several reports of Parthenium infestation in forest area and with a conservative estimate of only 1% area to be infested with Parthenium, it comes to about 0.68 million hectares. Thus in the light of above consideration and keeping in mind the Gidwani's (1975) estimation about 30 years back, in a very modest estimate, the present infestation of Parthenium in India is around 10 million hectares (National Research Centre for Weeds-Annual Reports).

In Other Countries

The global spread of Parthenium is given in Table 8. Parthenium is reported to be native to West Indies, tropical South and North America (Castex *et al.*, 1940; Rao, 1956;

Fernold, 1970; Arny, 1897). Its growth is also reported in the United States of America from Florida to Texas, North of Massachusetts, Pennsylvania, Ohio, Michigan, Illinois, Missouri and Kansas (Gleason and Cronquist, 1963; Fernold, 1970). It has also been reported from New Mexico (Fernandez, 1942). Minnesota (Mackoff and Dhal, 1951) and Louisiana (Ogden, 1957). Lately, it has been observed on ore piles stocked at Canton, port of Baltimore, Maryland, New Port and Virginia in Eastern United States (Reed, 1964). There are also reports of its presence in areas extending from Mexico up to Uruguay. In Argentina, it has occupied large areas and has also been spotted in the province of Cordoba during 1939 (Castex *et al.*, 1940). As per the report of Krishna Murthy *et al.* (1976a), the weed is observed in pure dense stands on fallow lands along the highways in Trinidad. In Queensland, the weed was first reported in Central Highlands near Twin hills during 1973 and subsequently in North Clermont. Despite a major campaign to contain the weed, it continues to spread and a recent report suggests that it could eventually become established in disturbed soil well in New South Wales (Dale, 1980). The weed became aggressive in a region within about 2 degrees north and south of the tropic of Capricorn. There have also been documents on its presence in South Africa, Islands of Mauritius, Rodriguez, Seychelles, Bourbon in the Indian Ocean and further in North Vietnam and in most places, the weed has become naturalized (Maheshwari, 1966; Vaid and Naithani, 1970; Maheshwari and Pandey, 1973).

Its occurrence is reported from the neighboring country of Bangla Desh though not in an alarming proportion. Enquiries made with the Sri Lankan scientists did not give any indication of its presence in Sri Lanka. The Malaysian weed scientists have realized its ill-effects and are very cautious to see that the weed does not enter their country as expressed by them to the author at the International Symposium on Biological Control in Tropical Agriculture, held at Kuala Lumpur, Malaysia from 25-30

Table 8. Global spread of Parthenium

Country	Provinces	References
USA	- Florida, Texas, North of Massachussetts	Castex *et al.* (1940)
	- Pennsylvania, Ohio, Michigan, Illinois, Missori and Kansas	Arny (1897) Fernod (1970)
	- Minnesota	Mackoff and Dahl (1951)
	- Louisiana	Ogden (1957)
	- Cantan, Baltimore, Maryland, New Port and Virginia	Arny (1897)
Mexico	- Throughout the country	Arny (1897)
Argentina	- Large areas, province of Cardoba	Castex *et al.* (1940)
Trinidad	- Large areas – major problem	Krishna Murthy *et al.* (1976a)
Guyana	- Large areas-major problem	Krishna Murthy *et al.* (1976a)
Jamaica	- Large areas-major problem	Krishna Murthy *et al.* (1976a)
Australia	- Queensland, Central Highlands and North Clermonth, New South Wales	Dale (1980)
South Africa	- Large areas	Maheshwari (1966)
		Vaid and Naithani (1970)
		Maheshwari and Pandey (1973)

Table 8. *Continued ....*

Country	Provinces	References
Mauritius	- Large areas	-do-
Rodriguez	- Large areas	-do-
Sychelles	- Large areas	-do-
Bourbon	- Large areas	-do-
North Vietnam	- Large areas	-do-
Bangladesh	- Large areas	Mahadeva-ppa (1996)

August, 1991. It is not also seen in Singapore and Indonesia as expressed by the scientists of the respective countries and the scientists are aware of its ill effects and therefore, are taking all precautions not to allow the weed to invade into their land.

During the months of January and February, 1988, the author had a chance to see the weed growing in the southern parts of USA, Mexico and Cuba. Parthenium was seen growing like any other plant and it did not dominate other plant species. It did not give an impression that it has invasive tendency. It needs to be established through a survey whether the prevalence of this plant is limited to the countries already listed or it is present in other countries too. Its presence may go unnoticed unless it dominates over other plant species or causes hazards to the human health and livestock. In fact, when the author observed in different countries, it was never a dominant species over the other naturally existing waste land species but grew in normal proportion like any other plant in that area. This probably is a settlement of a balanced biodiversity of that region.

As far in Australia, it has become widespread in grazing land from central Queensland to northern New South Wales. It causes direct losses to the grazing industry (about $A 14-18 million per annum) and is a human health hazard, causing allergic rhinitis and contact dermatitis. Starting in 1976, the Queensland Department of Lands (now Natural Resources, DNR) has had an on-going campaign to reduce the spread and impact of this public nuisance. However, the parthenium has nevertheless continued to increase and spread. Two biotypes of *Parthenium hysterophorus* L. have established in Australia as a result of two separate introductions from the USA. The first introduction occurred in south-east Queensland and the second in central Queensland. Nine plants from each of the biotypes were grown under a day/night temperature regime of 23/13°C and 14.5 h photoperiod in a plant growth cabinet for a period of five months. Plants from the central Queensland biotype had a higher dry weight production, earlier stem elongation, were taller; produced larger diaspores and had a lower percentage of filled seed than the plants from the south-east Queensland population. These differences in biology may offer an explanation why the central Queensland biotype is very aggressive while the south-east Queensland biotype is relatively well contained (Navie *et al.*, 1996). In Queensland, Parthenium plant commonly dominates cultivated and other disturbed areas, in addition to flood-prone pastures. The presence of Parthenium in cropped lands can almost double cultivation costs and restricts the sale and movement of contaminated produce. In 1990–91, a mail survey of beef producers was conducted in the most heavily infested region in central Queensland. Annual losses caused by this weed were found to be in the vicinity of $16.5 m. Losses comprised opportunity costs (*e.g.* reduced stock numbers and live weight gains), as well as additional production and control costs. Increased expenditure on research into parthenium control (especially biological control, for which research expenditure was approximately $350 000 during 1990–91) is thus warranted (Adkins *et al.,* 1996).

It is also evident that the entry of the weed into Sri Lanka has been multiple in time and space during the past two decades. Apart from the historical introduction of the weed by the IPKF and other Indian personnel, the commonest mode of recurrent entry of the Parthenium seed into Sri Lanka is apparently along with the import of condiments, *e.g.*, mustard, chilli, onion and onion seeds, from India. Considering the notoriety and behaviour of the Parthenium weed in other countries, such as India and Australia, its invasiveness and associated problems in Sri Lanka were imminent.

Five

HARMFUL EFFECTS OF PARTHENIUM

In general, Parthenium is a poisonous, pernicious, problematic, allergic and aggressive weed posing a serious threat to human beings, their livestock and flora. In India and Australia, this weed has been considered as one of the greatest source of dermatitis, asthma, nasal-dermal and Naso-bronchial types of diseases. Besides ill-effects, it also causes several other problems like blockage of common pathways and reduces the aesthetic values of parks, gardens and residential colonies. In agricultural fields, where only one crop is taken in a year, it grows profusely in fallow period with the occurrence of mild rains. Parthenium grows luxuriantly in orchards due to less frequent weeding in these areas. Invasion of parthenium in nearby vacant community lands and crop lands has reduced the suitable places for poor rural people for their natural calls. The effect of parthenium on man, animals and agriculture is given below.

When human beings frequently come in contact of this weed (Plate 27a), it may cause allergy, dermatitis, eczema, black spots blisters around eyes, burning, rings and blisters over skin, redness of skin, asthma etc. A major population does not feel any sensitization when they cone in contact with parthenium first time or for some time. But chances of getting sensitized to the weed

Ill-effects or parthenium in human beings

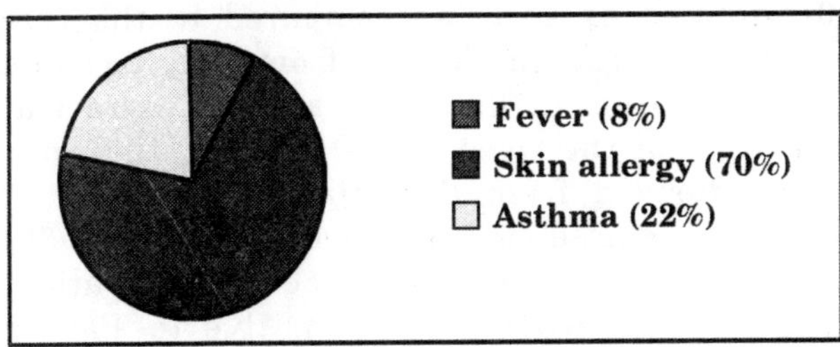

are high, when a person comes in regular contact for a period ranging from 4–15 months. Out of 56 weeds screened for their ability to cause dermatitis on 50 patients, Shelmire (1940) found that *Ambrosia eliator* (short ragweed) (23/50 cases), *Helenium microcephalum* (28/50), *H. tenuifolium* (27/50), *Lva angustifolia* (Marshelder) (18/50), *P. hysterophorus* (27/50) and *Santhium speciosum* (11/50) gave positive patch test reactions, most frequently. Likewise, Mackoff and Dhal (1951) observed positive reactions to wild feverfew, *P. hysterophorus* (16/25), *Franseria acanthicarpa* (15/25), *Ambrosia artemisiifolia* (22/25), *Lva xanthifolia* (15/25), *Xanthium* spp. (14/25), *Helenium autumnale* (20/25) and *Artemisia ludoviciana* (17/25). However, Mitchell *et al.* (1972) reported that about 34 species of 16 genera of the Helianthae tribe of chrysanthemum plants were found to cause allergic contact dermatitis. Allergic reactions are not always experienced with the first contact with the plant but may develop after a number of exposures. Once a reaction to parthenium weed develops, some people may also show similar reaction to other related plants (*e.g.* sunflower). The reaction can be so severe that people are forced to leave weed infested area. Subsequent patch test study by Lonkar *et al.* (1974) indicated that *P. hysterophorus* exhibited more frequent positive reactions on the patients as compared to ragweed, cocklebur (*Xanthium strumarium*), marshelder (*Lva* spp.), chrysanthemum etc. Further, patients showed more patch test reactions to parthenium leaf (50/50) and flower head (40/50), applied as such, as compared to the reactions either due to parthenin (derived from *P. hysterophous)* or to 1% in petroltum (27/50) or to acetone extract at 1/10 concentration of the weed (20/50). While, the compound abrasion (derived from *P. bipinnatifidum*) in, 1% in petrolatum, produced more positive reactions (33/50) than parthenin. Again, Shelmire (1940) observed a patient with Parthenium dermatitis to be sensitive to Parthenium extract 1:1, 00,000 but showed negative reactions to 1:10 extract of 68 other common weeds. Thus, a high degree of specificity may be observed in dermatitis caused from composite plants (Lonkar *et al.,* 1974). Though the weed is observed in varying intensities the allergic disorder due to this weed is not reported in West Indies (Krishna Murthy *et al.*, 1977). Inherited predispositions among the people in a country may be one of the reasons for the observed variations in reactions.

About **4%** of exposed males (farmers & field workers) acquire parthenium dermatitis (PD). The genetic or acquired tolerance in people of USA–may be due to exposure early in life. It affects almost exclusively men (40–60yr.) and rare in women. It does not affect children. The disease gets **activated in sun light**. In *Parthenium hysterophorus*–parthenolide, the principal allergen is an oil-soluble oleoresin present throughout the plant and pollen, responsible for PD parthenolide, an incomplete antigen (hapten) in the presence of sunlight (UV) combines with albumin in dermal part of skin, becomes complete antigen and causes photosensitive reaction resulting in photo dermatitis.

Parthenium dermatitis has various clinical patterns which include phytophotodermatitis, exfoliative dermatitis, exacerbation of atopic dermatitis, exacerbation of seborrhoeic dermatitis, and eyelid dermatitis. Common photodermatoses include, in some sensitized persons, even a slight physical contact with the weed may cause swelling and blisters eruption on the skin on the exposed part of the body. The affected skin may show vasiculation with exudation. When the dermatitis becomes chronic, it spreads to wrists, forearms and upper trunk. In the secondary stage impetiginization, fissuring of the skin and pigmentary changes take place.

- Idiopathic group: Polymorphic light eruption (PMLE), hydroa vacciniforme, actinic prurigo
- Photo allergic dermatitis: contactants, drugs, chemicals
- DNA repaired-defective disorders: *e.g.* xeroderma pigmentosum
- Dermatomes exacerbated by UV light: LE, CTCL, LP, rosacea, pemhigus erythematosus acne, atopic dermatitis.

Studies to detect cases of Parthenium dermatitis (Parthenium sensitivity) among different photodermatoses patients have been made and indicate hypersensitively to this weed (Tables 9 to 11).

First report of contact dermatitis due to this weed in USA was made by French (1930) from Texas and Rio-Grande valley. The number of cases seen in Texas and the seasonal eruption of the exposed skin surfaces happened to coincide with the growing season of this plant Khan and Grothaus (1936). Shemire (1939 and 1940) observed the attainment of acuteness of the disease due to *P. hysterophorus* inflicting more than 50% of the contact

Table 9. Sex Ratio of patients

Male	Female
30	41
42.25%	57.74%

Table 10. Age Ratio of patients

Age of patient	No. of Plants	Percentage
20-29 yrs	15	21.2
30-39 yrs	17	23.94
40-49 yrs	19	26.76
50-59 yrs	8	11.26
60-70 yrs	13	18.30
Total	71	

Table 11. Occupation of patients

Occupation	No. of Plants	Percentage
House wife	27	38
Teacher	4	5.6
Govt. Servant	14	19.71
Labourer/Farmer	11	15.49
Retired	8	11.26
Businessman	2	2.8
Unemployed	5	7
Total	71	100

dermatitis caused due to weeds especially in southern parts of the United States. He observed chronic lichenified eczema of the exposed skin surfaces which tended to be perennial owing to continuous exposure to the weed. Further, clinical confirmation was obtained through positive patch test with the leaf of Parthenium on four patients at Louisiana (Ogden, 1957). Even the related *P. argentatum* species had been reported to cause allergic contact dermatitis (Smith and Hughes, 1938).

Lonkar *et al.* (1974) reported that no major outbreak of weed dermatitis has been observed in Texas (Shelmire,

1940) and /Minnesota (Mackoff and Dhal, 1951) where Parthenium was found to be a major sensitizer in causing contact dermatitis. Though sufficient data on the epidemiological or ecological speculations are seldom available for this, they indicated that the residents of the United States possibly have a genetic or acquired tolerance. They also indicated that such tolerance or immunologic unresponsiveness might have been resulted from exposure to species of composites, especially Parthenium and ragweed during early life.

In Pune, Ranade (1976) detected *P. hysterophorus* as an etiological factor in causing eczematous dermatitis in 239 cases. He indicated that the pollen of this weed was the main cause for this eczematous dermatitis. In a further study he reported 96 cases of dermatitis due to this weed (Ranade, 1975 and 1976). In Poona, Lonkar and Jog (1972) observed dermatitis of the exposed skin surfaces in males engaged in agricultural work during 1965. They attributed this type of skin disease to occupational exposure to Parthenium. Both Lonkar and Ranade have indicated that the allergic people are very much prone to the development of eczematous dermatitis on contact with this weed. There are also reports on the development of dermatitis in the local people while cutting this plant for firewood in Bihar (Suresh Chandra, 1973). Lonkar *et al.* (1974) indicated the attainment of epidemic status by affecting nearly 500 persons, predominantly of the male sex and has observed even ladies developing this symptom on exposure to this plant in Dharwad, Karnataka. Similarly, at Coimbatore, both men and women developed allergic symptoms on repeated contact with the weed. Further, women developed cracks all over the sole, especially in the heel (Sundara Rajulu and Gowri, 1976). There are evidences of allergic papules in school boys when they had volunteered for uprooting Parthenium in Hassan (Krishna Murthy *et al.,* 1975). Further, it has been observed that chances of getting sensitized to the weed are high, when a person comes in constant contact with the weed for a period ranging from 3 to 12 months (Subba Rao *et al.*, 1976).

The progressive symptoms observed on patients as confirmed clinically by Lonkar *et al.* (1974) are as follows.

During early acute state, acute eczematous dermatitis affected the face, antecubital and popliteal fossae, the V of the neck and hands (Plate 9a, b). The skin showed vesiculation with exudation and intense pruritus. There was uniform failure of response to topical corticosteroid drugs and the dermatitis progressed to chronic dermatitis with lichenification and spread to affect the wrists forearms, upper trunk and in some instances become partially generalized or universal. In chronic stage, systemic corticosteroid drugs were ineffective. The secondary changes included impetiginization, fissuring of the skin and various pigmentary changes. Both post inflammatory hyper pigmentation and hypo pigmentation were seen, the former predominating in most instances. The course of the disease was progressive with exacerbations during the growing seasons of the plant and was uninfluenced by therapy. However, in cases where follow up was possible, complete remission was observed following a transfer of residence to geographic area not infested with the weed, only pigmentary changes persisted after such a change of residence. The patients with generalized or universal dermatitis, particularly if caecum exudation and impertiginization were severe, became ill. They suffered fatigue associated with intense natural pruritus, malaise and loss of weight. About 12 deaths occurred in severely affected patients. Intecurrent infection-cutaneous and pulmonary, were considered to be responsible for death. Associate features of severe dermatitis were loss of scalp and body hair the nails showed shiny polished surfaces or transverse ridging but are rarely shed (Krishna Murthy *et al.,* 1975). The skin allergic reactions on face and back of the persons were observed by the doctors in Dharwad and Gadag (Plates 10, 11 & 12a).

The author further observed the severity of the disease on males as compared to females to the extent of 10:1 and the disease was not observed before puberty. The author also indicated chances of spontaneous recovery from repeated topical applications despite continued exposure to the allergens.

Ranade (1975 and 1976) reported that he had developed a vaccine for desensitization of people suffering from the disease.

The response of the patients to the vaccine has been provided by him. Earlier, steroids were used as anti-histaminics had no effect. He has treated more than 400 cases with gratifying results. The injections were given every day in increasing doses and the course was completed by 20 days. The results of hyposensitisation lasted for varying periods, not lasting more than 6 months in farmers and laborers and for more than a year or two in city dwellers. During early stages of actuate oozing, the patients responded to vaccine miraculously. In chronic lichenified cases, itching subsided with 8–10 injections but lichenification particularly of localized nature did not disappear. Diffuse infiltrative type clears completely in about 15 to 20 days. He also indicated more chances of relapses. Nevertheless, the booster doses once a year before monsoon (June–July) kept the patients free throughout the growth of the weed during the year. According to him, about 8% of the patients he treated were free from the trouble as observed over 5 to 6 years.

While this being so, Kelkar (1970) from Poona has observed that though Parthenium possessed certain activity as a depressant to the central nervous system, it is too toxic and warrants further investigations on human beings.

The author has personal knowledge of 55 individuals in Bangalore city who are suffering from Parthenium allergy. In one case, a lady aged forty years was sweating out water from her skin to the extent of 1.0 to 1.5 litre per day during peak season of Parthenium growth. It was not so serious when she lived in a Parthenium free environment or when the Parthenium population existed in small proportion in the neighboring fields. In another case, a man aged 47 years sold away his house in an extension area infested with parthenium where he was suffering from Parthenium allergy and went back to his old residence in down town area where there were no Parthenium plants and hence he could get relief. This calls for a study to find out the level of permissible parthenium population in an area, safe to human beings or cattle.

Parthenium is reported to be responsible for largest number of airborne contact dermatitis in India. Almost every part of the plant except root is reactive. Any person sensitive to Parthenium may develop symptoms by coming in contact with the pollens and trichomes, which are available in the atmosphere in the Parthenium infested area. A significant proportion of bronchial asthma patients was sensitized to *P. hysterophorus* (Suresh *et al.,* 1994). Fisher (1996) described the incidence and severity of contact dermatitis in humans caused by Parthenium in India. It was argued to be the first reported instance of an imported sensitizer causing allergic dermatitis in thousands of people. Kologi *et al.* (1997) also reported the dermatological hazards in human beings (National Research Centre for Weeds-Annual Reports). Parthenin, an incomplete antigen heptain in presence of sunlight (ultraviolet rays) combines with albumin in the dermal part of the skin, becomes complete antigen and causes photosensitive reaction. This is called photophyto dermatitis. It can also cause the contract dermatitis.

By continuous contact to the plant for 8–10 years, patients complained itching and later developed rashes and boils. The reaction was mainly over the sun exposed area.

Effects on Livestock

Parthenium is also reported to have harmful effect on livestock. It is a threat to livestock as it reduces the availability of fodder in pastures and is toxic when ingested. The weed is not palatable to livestock due to its irritating odour, taste and presence of trichomes. But sometimes-hungry cattle eat the parthenium. Goats and stray cows are often seen eating parthenium in summer and rainy season. In animals also, parthenium was found to cause clinical signs such as salivation, onset of diarrhoea when calves were fed on Parthenium, anorexia, pruritus, alopecia and dermatitis (Plate 12b), and gastro-intestinal irritation resulting in diarrhoea. Anorexia developed but it declined when calves were accustomed to the feed. Parthenium caused itching, alopecia, and dermatitis on the face, muzzle, neck, eyes, thorax, abdomen and brisket region in calves (Fisher, 1996). In cattle, due to Parthenium contact, there may be loss of hair and marked depigmentation of skin. The bitter and reduced milk yields have been reported in buffaloes and goats, fed on grass mixed with parthenium (Krishnamurthy *et al.*, 1977) (National Research Centre for Weeds-Annual Reports). Though cattle do not eat Parthenium, its effects were observed on them when they

Ill-effects of Parthenium in animals

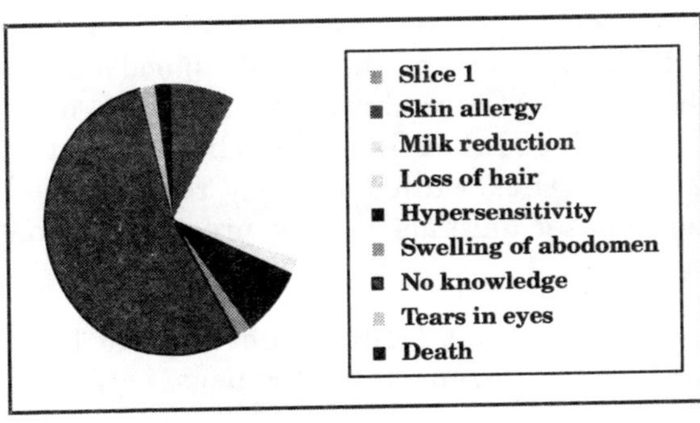

Years Parthenium invasion

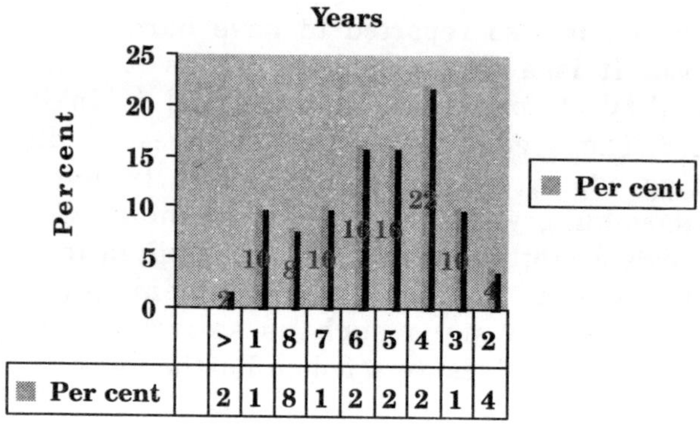

walk by or graze through patches of this weed. Such cattle had inflamed udder and subsequently suffered from fever and rashes (Krishna Murthy *et al.*, 1977). In a recent survey, they observed cows (milking or otherwise), buffaloes, and poultry birds pecking the flowers. Their main concern was about the probable contamination of the milk with the active principle, parthenin, when the livestock eat the weed. Sundara Rajulu and Gowri (1976) studied the harmful effect due to repeated contact with Parthenium on a young rabbit by letting it free to move about in densely infested weed at Coimbatore. The rabbit exhibited hypersensitivity by showing restlessness from the eighth day onwards. After 3 days, natural falling of hairs from the dorsal region of the neck and back was observed. By the 12th day, small boils erupted all over the neck and lateral sides of the trunk, blood oozed out on the 13th day and the rabbit died on the 17th day. It was, therefore, concluded that the weed is hazardous to animals too. However, the studies in this regard need to be extended to other animals also for precise information on hazardousness.

It is reported that feeding the weed to buffalo and bull calves at different levels causes both acute and chronic forms of toxicity. Ulcerations were caused both in the mouth and digestive tract. Autopsy of the dead animals showed punched out ulcers on the esophagus and the abysmal folds. Histopathology of the kidney and liver revealed degenerative changes and necrosis (Subba Rao *et al.,* 1976).

It was reported that "labeled parthenin" was found to be excreted in the milk when administered to lactating guinea pigs, rabbits and a cow. Consumption of milk from the Parthenium fed live stock could be hazardous to man.

With the information available from other countries and from our own surroundings, it is now certain that this is not a humble weed. Though more medical studies are needed to draw definite conclusion, none would doubt its health hazards. In the light of these discussions, it is clear that all-round efforts should be made to keep the weed under control (Krishna Murthy *et al.*, 1977). Effect of Parthenium on human health and livestock is given in **Table 12.**

Though Parthenium is normally not grazed by the cattle, as it is not palatable due to its irritating odour, bad taste and presence of trichomes, however, stray cattle often are forced to feed on this weed during periods of fodder scarcity thereby resulting in impairment of both quality and quantity of milk. Parthenium toxicity involves alopecia, and dermatitis on the face, muzzle, neck, eyes, thorax, abdomen and brisket region and causes salivation, anorexia, diarrhoea and death in extreme cases (Kandhane *et al.*, 1992).

Effects in agricultural, horticultural and forestry crops

Many phenolics are said to be responsible for allelopathic impact of Parthenium on certain crops (Kanchan, 1975). It has been established that growth inhibitors are released

Table 12. Effect of Parthenium on human health and livestock

Disorder	Reaction	References
Contact dermatitis	Seasonal eruption of the exposed skin surface	French (1930) Khan and Grothaus (1936) Shelmire (1939, 1940)
Eczema	Chronic lichensified eczema of the exposed skin surfaces	Orden (1957) Smith and Hughes (1939)
Eczematoid dermatitis	Skin eruptions and itching	Ranade (1975, 1976)
Eczematoid dermatitis	-do-	Suresh Chandra (1973)
Dermatitis	-do-	Lonkar *et al.* (1974)
Allergic reactions	Cracks all over the sole	Sundara Rajulu and Gowri (1976)
Allergic papules	Sore throat, bubbles in the mouth	Krishna Murthy *et al.* (1975) Subba Rao *et al.* (1976)
Fatigueness	General weakness, skin eruptions	Mahadevappa (1996)
Severe dermatitis	Loss of scalp, body hair, ridging on nails	Krishna Murthy *et al.* (1975)
Fever in cows	inflamed udder and rashes	Krishna Murhty *et al.* (1975)

Table 12. *Continued* ......

Disorder	Reaction	References
Hypersensitivity in Rabbit	Restlessness, natural falling of hairs from the dorsal region of the neck and back, small boils and oozing of boils	Sundara Rajulu and Gowri (1976)
Ulcerations in buffaloes, horses, donkeys, sheep and goats	Acute and chronic toxicity, ulcers both in the mouth and digestive tract, oesophagus and abnormal folds, necrosis of kidney and liver	Subba Rao *et al.* (1976)

from this plant into the soil through leaching, exudation of roots and during decay of residues (Kanchan and Jayachandra, 1979a, 1979b, 1980). These inhibitors, identified as sesquiterpene lactones (mainly parthenin) and phenolics, are produced continuously by healthy plants, remain active in the soil for up to 30 days and suppress local vegetation. As a result, parthenium can be seen in dense, pure stands extending over several hectares, threatening nature's diversity. It has also been reported to invade cultivated fields of most cereals, vegetables, oilseeds and plantation crops, posing a serious threat to agriculture and horticulture. It is reported to cause loss to field crops like ladies finger, tomato, beans, capsicum, maize and brinjal etc. A loss in yield of crop up to 40% (Khosla and Sobti, 1979) and reduction in forage production up to 90% have been reported in grazing lands in the state of Maharashtra (Vartak, 1968).

Ecological problems faced in India due to invasion of exotic species including *P. hysterophorus* into grazing and shrub lands have also been documented. This invasion of exotic species is the most important problem since it is leading to a loss of palatable and other economically important species. A change in the nutrient composition of the soil has also been observed. The threat to existing plant species and grazing lands was a matter of concern for the environmental management in Garhwal region of Uttaranchal (Rajwar *et al.,* 1998). Kumar and Rohatgi (1999) feared the role of invasive weeds in changing floristic diversity. They were of the opinion that some invasive weeds including Parthenium dominate fields, forests and wastelands, occupying almost the whole area, with little or no growth of any other species. The invasion of Parthenium has threatened the forest biodiversity in sal (*Shorea robusta*) forest of Madhya Pradesh (Pandey and Saini, 2002) (National Research Centre for Weeds-Annual Reports).

It was predicted that, in Pakistan, the population of many common medicinal plants growing in the wastelands of Islamabad may rapidly decline due to the aggressive colonization by Parthenium weed (Asad Shabbir and Rukshsana Bajwat, 2005).

Parthenium pollen and its effects

Ever since the Parthenium plant started invading newer areas in high population, its multitude of ill-effects on human beings and cattle have been reported in plenty. But no systematic studies have been made to know which part of the plant is dangerous to human and animal life. A few studies carried out using different plant parts are reviewed below.

In atmospheric pollen survey carried out in Delhi during 1958–59, Dua and Shivpuri (1969) recorded pollens of many anemophilous species. It was dominated by grasses and other weeds during August to October or early November; but they could not observe the pollen of Parthenium. While, in a subsequent study on the unknown aero allergenic pollens of Delhi Metropolitan area during 1965–66, Kartar Singh and Shivpuri (1971) observed that the atmosphere was never free of pollen but the highest peaks were noticed during March–April/mid May (spring or early summer) and September–October (late rainy season and autumn). However, Parthenium pollens were observed in the atmosphere throughout the year but its prevalence was rated as 'common' though there were two other higher grades as "very common" and "abundant" in the classification. They also indicated that this amphiphilous group plays in important role as allergens affecting 30% of the population. Subsequent study at Delhi indicated that the pollen of Parthenium showed marked positive skin reactions (2+ and 3+) in 14 out of 50 patients, while only 2 out of 25 healthy normal (8%) showed 2+ reactions (Singh *et al.,* 1974). They observed that flower extracts stand next to pollen in allergenicity. The flower extract showed positive skin reactions in 3 (6%) out of 50 patients, closely followed by leaf extract (2 out of 50).

Agashe and Prathiba Vinay (1975) observed the atmosphere of the Bangalore city with full of pollen pollutants. The authors observed the pollen morphology of Parthenium in all aeroplynological studies of Bangalore city. Studies require to be taken up to find out the presence of the pollen grains (Plate 13) of this weed in the atmosphere, and the different types of allergies caused by the pollen. Nevertheless, there is a need for study of pollen, the pubescent hairs and other parts of Parthenium plant in greater detail. A detailed survey of atmospheric pollen conducted by Subba Rao (1984) in Bangalore indicated varying amount of pollen grains either singly or in clumps through out the year. However, in the months of June and August, the Parthenium pollen was reported to occur in abundance.

In general, pollen activity causing allergy has been considered to be well known. In a study carried out at Jaipur on the allergenic disorder caused by allergens on 100 people, it has been indicated that nearly 50% of the population developed asthma, 16 to17% with urinary and rhinitis, while only 4% of the people developed dermatitis (Krishna Murthy *et al.,* 1977). The people in the age group of 21 to 30 years suffered largely (by 21%) followed by 11 to 20 years (by 11%) (Kasliwal *et al.*, 1961). As the pollen of Parthenium is anemophilous of extensive dispersion (Castex *et al.*, 1940), it can be an important source of allergens to the susceptible people (Kartar Singh and Shivpuri, 1971). Though the reports are available on the indication of asthmatic allergy, they need scientific confirmation. The pollen of Parthenium has been observed to cause allergic rhinitis in Argentina (Castex *et al.*, 1940) and Mexico (Fernandez, 1942) but not in North America (Wood House, 1971). The possible reason put forth for this was that the pollen is not liberated as airborne individual particles and will not be wind borne to any large extent except in conditions of summer wind streams. Besides, the aggregation of particles in the gummy part of the flower head was observed (Khan, 1924). Based on this, Khan and Grothaus (1936) emphasized that the plants were unimportant in causing high fever in Texas. However, Lonkar *et al.* (1974) indicated the necessity of such confirmatory traits under Indian conditions. In a pollen study at Bangalore, Kanchan and Jayachandra (1976a) observed parthenium to produce an average of 624 million pollens per plant and these were carried to distant places in clusters of 600–800 grains each.

The pollen is also reported to have allelopathic activity and has inhibited development of fruits in brinjal, tomato, chilli when artificially dusted on stigmatic surface of these plants. Further, accumulation of 100–150 clusters of Parthenium pollen on floral parts of maize caused 50% reduction in grain filling. In nature, heavy deposition of pollen of this weed on floral parts of *Crotolaria pellida and Desmodium hectrocaqron* has also shown to result in poor fruit setting (Kanchan and Jayachandra, 1976a). This inhibitory effect of pollen on the growth of a fungus or pollen of other species is referred to as allelopathic mechanism, in which species compete against others by releasing growth inhibiting chemicals (Whittaker and Feeny, 1971; Kanchan, 1975a).

Pollen allergy studies in wasteland weeds

Effective control of Parthenium by growing *Cassia sericea* has already been established by earlier workers. A few other plant species have also been reported to control spread of Parthenium. To clear the doubts expressed about allergenic effects of *Cassia sericea* and such other plant species on human beings, a study was conducted to evaluate pollen allergenicity by skin testing with some wasteland weeds namely, *Cassia sericea, C. tora, C. auriculata, C. occidentalis. Sida spinosa, Amaranthus spinosus, Mirabilis jalapa, Croton sparciflorus* and *Ipomoea* spp. Despite variation in the size of the pollen grains of different species studied, no correlation was

Table 13. Phenology, pollination type, pollen incidence, pollen morphology and pollen size of wasteland Weeds

Family	Species	Mode of polli-nation	Flowering period	Pollen Size (m)	Pollen morphology	
					Aperture ornamentation	Exine
Asteraceae	*Parthenium hysterophorus*	AN	All round the year	17.14	3-Colparate	Spinulose
Fabaceae	*Cassia sericea*	AM	August-November	36.24	3-Zono-colparate	Thin, psilate
Fabaceae	*C. tora*	AM	August-November	44.67	3-Zono-colparate	Thin, psilate
Fabaceae	*C. occidentalis*	AM	August-October	40.30	3-Zono-colparate	Thin, psilate
Malvaceae	*Sida spinosa*	AM	August-November	89.20	Pantaporate	Thick walled spinulose
Convolvulaceae	*Ipomoea* sp.	EN	July-October	116.10	Pantaporate	Thin, psilate
Amaranthaceae	*Amaranthus spinosus*	AN	August-November	25.45	Pantaporate	Spinulose
Nyctaginaceae	*Mirabilis jalapa*	EN	August-December	174.40	Pantaporate Conspicuous strata	Thick exine spinulose
Euphorbiaceae	*Croton sparci-florus*	AN	All round the year	42.47	Inaporturate	Finely reticulate

AM=Amphilous; AN = Anemophilous; EN = Entemophilous. *Source*: Nalini *et al.* (1997)

observed between pollen morphology and allergenicity. The overall incidence of allergy with respect to different parameters like sex, age, and weight and clinical history was very low for all the species. Out of nine species tested, only three species namely, *Amaranthus spinosus*, *Mirabilis jalapa* and *Cassia occidentalis* were found to possess some allergenicity and the remaining six species exhibited negligible allergenicity. *C. sericea* can thus be very safely used along with other five species for suppressing the growth of Parthenium **(Table 13).**

It was concluded that Parthenium as well as Xanthium plant act as a precipitating factor in 18% of Photo-dermatoses patients. Females are more vulnerable to Parthenium sensitivity. However, inclusion of a larger number of patients in next two years for this project will throw more light on the definite role of Parthenium and other composite plants in various photodermatoses. It was felt that the role of ultra violet light in exaggeration of symptom and sign of Parthenium dermatitis leading to development of different secondary photodermatoses including chronic actinic dermatitis has to be kept in mind.

Six

METHODS OF CONTROL

Ever since Parthenium assumed a menacing proportion in different parts of the country, several methods are being recommended in containing the growth of this plant. When individual Parthenium plants are easily killed by weedicides or by any means rapid regeneration from seed follows. Therefore no single method appears to be satisfactory, as each method tried during the period from mid sixties to early nineties suffered with one or the other limitation such as inefficiency, high cost, impracticability, polluting the environment, only temporary relief, fear of allergic reaction etc. However, the integrated approach recommended in the late eighties seems to be promising. The various methods *viz* i) manual, ii) chemical, iii) cultural, iv) utilization, v) biological and vi) integrated methods together with their merits and demerits are discussed in this chapter (Mahadevappa and Gautam, 2006). Further, people need to be aware of how Parthenium spreads in order to take preventive measure to contain it. Therefore, field days, awareness camps and meetings with interested landholders are to be regularly carried out. Further, it is very important to educate people about how the plant looks like and the hazards associated with it. Education to students as well as workers from construction companies who may work with contaminated machinery around is also very important. Most important is to pass on the current knowledge obtained through research carried out on Parthenium to the youth in order to educate them for managing the area in future.

Awareness raising activities have a special significance in Parthenium management. Therefore it is necessary to understand public awareness on Parthenium and its biological agents **(Table 14)** for developing a comprehensive intended control programme.

Manual Method

Removing Parthenium plants by hand has been employed for the past several years but with disappointing results. In small areas and isolated pockets such as flower beds, lawns, kitchen gardens and in intensively cultivated agricultural fields, hand weeding can be really effective and should be preferred. The plants should be uprooted (not cut or broken) before flowering and burnt or composted. If uprooting is done after the flowering stage, the pulled out plants are to be burnt or buried without transporting to far off places, to avoid seed dispersal. Manual method also finds a place in the integrated approach in order to achieve quick results. When biocontrol methods are adopted, there will always be certain proportion of plants left unchecked which is typical of any biocontrol method. Physical removal of such remnants (left out plants) will supplement the biocontrol efforts in checking the growth of Parthenium to a satisfactory level. But manual method is neither economical nor practicable in vast areas with heavy infestation. In limited situations, the method can work as a compliment. It is to be ensured that uprooting the plants is carried out by engaging persons insensitive to Parthenium allergy even in such cases, with safety measures such as wearing of hand gloves and nose covers

Table 14. The level of infestation and public awareness of Parthenium in the India

State	Infestation level *	Public awareness	Awareness about Mexican beetle	Mexican beetle presence	Mexican beetle impact
Andhra Pradesh	3	4	1	2	1
Bihar	4	4	0	0	0
Chhattisgarh	4	4	0	2	0
Gujarat	3	2	0	0	0
Haryana	4	4	0	4	1
Himachal Pradesh	2	4	0	2	0
Jharkhand	3	3	0	0	0
Karnataka	4	5	2	4	2
Kerala	1	2	0	1	0
Madhya Pradesh	4	4	2	4	2
Maharashtra	4	4	1	2	0
NE States	2	2	0	0	0
Orissa	2	1	0	0	0
Rajasthan	3	4	0	0	0
Tamil Nadu	3	4	2	3	1
Uttar Pradesh	4	4	1	3	2
West Bengal	2	5	0	0	0

*Rating: 0–lowest, 5–highest

are necessary. It is also necessary to train them to see that the plants are removed till the tip of the root. It is commonly observed that the stems break at the ground level, remain dormant in "rosette" form and sprout when some moisture becomes available in the soil. However, it should be noted that the relief expected from this method alone is only temporary and needs to be repeated as and when the weed appears. As per Australian experience, mechanical control methods such as grading, slashing and ploughing for controlling Parthenium are not effective. Manual cutting like mowing or slashing results in rapid regeneration of plants and subsequently allows quick flowering with abundant seed production.

Chemical Method

Several selective herbicides, commercial weedicides, soap water and salt water have been tried in experimental plots and subsequently under actual Parthenium infested fields. Chemicals like 2, 4-D sodium salt, arsenate compounds, paraquat, bromacil, glyhosate or sodium chloride in recommended dose kill the standing Parthenium populations and to some extent suppress immediate germination of the seeds deposited in the soil (Kasasian and Seeyave, 1969; Sukhda, 1975; Krishna Murthy *et al.,* 1977; Muniyappa, 1980). Immediate dying of Parthenium plants has been observed when a chemical named, Killer – 700 was sprayed to the established Parthenium plants. In cropped areas, herbicides have been recommended in different crops to control Parthenium in respective crops. Here, care needs to be taken to avoid spray drift reaching adjoining sensitive crops.

Fernoxone sprayed (80% sodium salt of 2, 4-D) at 5.0 kg/ha as post-emergent spray gave complete control of Parthenium at all stages of its growth within a period of 20 days after spray. In Karnataka (India), the following spray schedule has been recommended in non-cropped areas. Young Parthenium seedlings before flowering should be sprayed with 2, 4-D sodium salt @ kg/ha or MCPA @ 5 kg/ha or MSMA @ 5 litre/ha in 1000 litres of water. For grown up Parthenium plants, spray 2, 4-D sodim salt @ 2.5 kg + MSMA @ 5 litres/ha or 2, 4-D acinal @ 7.5 kg/ha + teepol @ 2.5 litre/ha or urea @ 5 kg/ha or MSMA @ 20 litre/ha in 1000 litres of water.

Herbicidal effect of chemicals does not last long it ends with only one flush of germination. Sometimes the plants so suppressed by chemicals have regenerated after remaining dormant for a few days. Thus, killing of established plants clear the way early only for the next flushes of Parthenium plants to emerge. Also chemical treatment repeatedly can kill the existing plants but cannot prevent the entry of seeds getting deposited from outside. The remnant seeds found in several thousands per every square meter of the invaded land as well as newly deposited seeds are always ready for germination with slight moisture becoming available to them in the soil medium. Because of continuous seed production without any interval in the entire calendar year, reinvasion of the areas can hardly be avoided unless the seed source itself is checked. Since Parthenium is more a waste land weed, there will be hardly anybody to invest on chemical treatment just for temporary relief. Even in the agricultural production the chemicals are found expensive and not within the reach of an average farmer. Further, repeated applications either of common salt, soap water or any herbicide will affect other beneficial plants, besides changing soil characters and polluting the environment due to cumulative residual effects. Even handling of chemicals needs expertise, training of skilled workers and adequate caution throughout the process of treatment. Hence, chemical method can not be relied upon even for a reasonable period of freedom from Parthenium. Nevertheless, there may arise situations where the vast stretches of already invaded Parthenium plots require being relieved of it immediately for emergencies wherein chemical control method only will have to be resorted to with the knowledge that it gives only temporary control and does not ensure a longer Parthenium free period. Further, in the integrated approach also, at the time of sowing and initial establishment of *Cassia sericea* and other plant enemies and one or two selective herbicide sprays would speed up the process of Parthenium replacement.

Cultural Method

Some researchers have advocated cultural methods by growing competitive crops to suppress Parthenium but, as Parthenium is mainly a wasteland plant the scope of this practice is limited to specific situations such as orchards, crop fields etc.

Utilization

The large-scale utilization of Parthenium can also be one of the effective methods as Parthenium has been well documented for its insecticidal, nematicidal and herbicidal properties (Gajendra and Gopalan, 1982; Bala *et al.*, 1986; Ramaswamy, 1997), besides oxalic acid (Mane *et al.*, 1986) and biogas production (Gunasheelan, 1987). It is also advocated to use abundant Parthenium biomass to make paper (Taneja *et al.*, 1997) and compost (Shukla and Singh, 1999). Parthenium can also be used to make compost and vermicompost. It has been observed that if compost is prepared through NADEP method, a lot of Parthenium seeds remain viable which may germinate in the crop field and may become a source for further spread of the weed. During a preliminary study, it was found that compost prepared by mixing Parthenium with dung slurry, soil and urea in layers in conventional pit in aerobic conditions, (provided with the help of wire mesh or chimney method) and removed after 4 months, was good to digest the Parthenium seed also and compost quality was superior to that of FYM. But, so far no effective technology for its utilization is developed hence above findings have remained of academic interest only.

Composting is also suggested as one of the methods to control Parthenium. The plant reaches 50% of seed setting during flowering. The plant left as such in the same area acts as a seed bank because of its higher seed production capacity and extended dormancy period. Hence composting is recommended, as the seeds lose their viability due to the higher temperature during composting. The suggested method is to cut the Parthenium plants into small pieces by using knife or chaff cutter and spread the material on the ground to a thickness of 10 cm layer. Over this *Tricoderma viridi* has to be spread and at 0.5 % urea solution sprayed (generally 5 kg urea/ton of plant material) (The Hindu, Thursday, Dec. 04, 2003).

Application of Parthenium or Chromolaena as green manure or compost in combination with recommended dose of fertilizer increased the status of available nutrients also and higher level of residual nutrient status in soil which might help the succeeding crop.

Parthenium composting

In integrated approach, mechanical, chemical, biological and cultural methods are mainly used for control of Parthenium. But none offers complete eradication due to various factors (Tripati *et al.*, 1991). Generally, the problematic weeds are manually collected and then disposed off or burnt. Studies to utilize weeds efficiently are lacking and efficient utilization of the weed through composting by using *Trichoderma viridae* as a culture, which not only manages the weed but also supplements the nutrient requirement to crops.

This sequence of layers is repeated up to a metre high, then plastering should be done with mud/clay soil.

Keep the moisture level at 50–60%. After two weeks, a thorough mixing has to be given.

The compost will be ready for field application after 40–45 days. It is a good source of nutrient and helps to maintain soil properties through aggregate formation. The 'parthenin' content acts as a growth regulator.

The Forest Research Institute in Dehra Dun claims to have evolved a high-value end use of it in the form of hand made paper, false ceilings, partitioning, table tops, cabinets and fiber boards. It is ascertained that the end products are free of allergic behavior for which Parthenium weed is notoriously known for. The commercial viability of these products is yet to be tested. This aspect calls for a serious consideration at government and/or NGO level to know its suitability for small scale and cottage industries.

Biological Method

Biological control is the intentional manipulation of natural enemies by man for the purpose of controlling harmful plants. Biological control of exotic terrestrial and aquatic weeds is especially attractive as wasteland, community land, rivers and lakes are sensitive ecosystems, important to wildlife and human heath. Biological control seldom means complete eradication of the unwanted organism, but rather maintaining its population at lower than average that would occur in the absence of the bio-control agent (National Research Centre for Weeds-Annual Reports).

Biological control is the use of living organisms to control unwanted animals or plants. A natural enemy such as parasite or predator or pathogen is introduced into the environment of pests or if already present is encouraged to become more effective in maintaining the number of pest organisms below the level of economic damage to crops or danger to human health. Biological control agents should be used as one tool in a combined management strategy which also incorporates herbicide and mechanical control of the several methods of pest control in crop production as well as in the living environment, biological approach is considered to be relatively less expensive and free from harmful side effects to the environment and the organisms habituated in the natural course. Insect pests, undesirable plants and many diseases were successfully controlled biologically in the past. Hence, biological control method is acclaimed as eco-friendly, acceptable, least expensive and sustainable. At present, much emphasis has been given through this approach to control Parthenium, although, like any other examples this is also beset with some limitations. Types of biocontrol agents used for Parthenium suppression are competitive plants, insects, fungi, nematodes, snails, slugs, and pathogens. So far success has been more with competitive plants in India followed by insects, whereas, in Australia insect agents have played very effective role. Some experiences are detailed below.

Parthenium control through biological agents is not as simple as in other successful cases because, Parthenium is hardy plant with exceptional adaptability to a wide range of ecosystems. The successful cases are the only ones involving the host organism having very specific adaptability. Use of more than one agent in an integrated approach is essential in suppressing this obnoxious weed. The research on the biological control has revealed that, at this point, there are both plant and insect enemies that can contain the growth of this weed. Use of biological agents such as virus and mycoplasma like organisms (MLOs) is still in the experimental stage. Eight species of exotic insects have been introduced into Australia for the biological control of Parthenium. The moth *Epiblema strenuana*, is exerting significant control on the plant.

Scenario of various biological enemies – **Biocontrol Agents** – is discussed below:

a) Botanical agents

There are two approaches in using plant enemies in the control of Parthenium. One is maintenance of naturally occurring biodiversity and the other is planting scientifically selected and proven group of plant species in target areas.

Maintenance of natural biodiversity : Protection of the naturally existing flora forms one of the three biocontrol aspects suggested in the integrated Parthenium Management. A botanical survey in relation to Parthenium control across the country has revealed an interesting factor that the Parthenium can not penetrate into areas where the natural floras have not been disturbed (Mamatha and Mahadevappa, 1988) (Plate 13). Wherever there is indiscriminate destruction of naturally existing plant species, the chances of Parthenium proliferation are more. For instance, in areas around cities and urban pockets like Hubli city (Karnataka) where most of the waste land plant species were destroyed in the process of clearing Parthenium, Parthenium continued to perpetuate, whereas in Dharwad city where no such Parthenium control measures were undertaken, Parthenium density is very less; because, *Cassia sericea, Cassia tora* and other waste land species have been preventing the entry and spread of Parthenium as can be seen now.

In Maharastra, *Styloxanthes scabra* Vogen has been found to compete with Parthenium through allelopathic effect. This has been confirmed by the field observation made by the Officers of the Department of Forests, Karnataka (A.S. Sadashivaiah – personal communication, 1991). According to Maheshwari (1996), the only other plant that competed with Parthenium was *Xanthium strumanium* – an allergic weed.

An experiment carried out in the University of Agricultural Sciences, Bangalore revealed that Parthenium seeds when sown on (i) barren (vegetation cleared) land and (ii) vegetation covered lands, could establish to the extent of 100% in the former and only to the extent of 13.8% in the latter (Mahadevappa and Ramaiah, 1988). Several other plant species were also identified as having similar impact but with varying degrees **(Table 15)** (Plates 14 to 25). The strongest ones among the so listed species in the order are (i) *Cassia*

Table 15. The plant species showing allelopathic effect (interference) on Parthenium growth

Plant	**Family**	**Level of Suppression**
Cassia sericea SW.	Leguminosae	Very high
Tephrosia purpurea Pers.	Leguminosae	High
Styloxanthes scabra Vogen	*Leguminosae*	High
Croton sparciflorus Morong	Euphorbiaceae	High
Hyptis suaveolens Poit.	Lamiaceae	High
Amaranthus spinosus L.	Amaranthacea	High
Sida spinosa L.	Malvaceae	High
Cassia occidentalis L.	Leguminosae	Moderate
Cassia tometosa L.	Leguminosae	Moderate
Cassia obtusifolia L.	Leguminosae	Moderate
Cassia tora L.	Leguminosae	Moderate
Cassia auriculata L.	Leguminosae	Moderate
Ipomoea muricata Jacq.	Convolvulaceae	Physical Supression
Ricinus communis L.	Convolvulaceae	Physical Supression
Ipomoea carnea Jacq.	Convolvulaceae	Physical Supression

sericea, (ii) *Tephrosia purpurea,* (iii) *Stylosanthes scabra,* (iv) *Croton sparciflorus,* (v) *Hyptis* spp., (vi) *Cassia tora* and (vii) *Amaranthus spinosus*. The author has observed dense growth of Parthenium in residential and industrial areas but almost negligent in lands away from such areas. Further, Parthenium growth is very intensive in places where new construction (like residential extensions in cities and towns, road and dam constructions etc.,) is going on. All these observations lead to the conclusion that maintenance of biodiversity, that is natural flora wherever possible, is important to check Parthenium entry/invasion and its growth as well.

Kauraw *et al.* (1997) reported that marigold grown in association with Parthenium reduced the height and root length of Parthenium. In the next generation, the population of Parthenium was reduced by 85 to 100 %. Thiophene compounds already reported from *Tagetus patula* L. roots seem to be phytotoxic to Parthenium. Regenerated plants have very poor root system and poor seedling development which even could not flower. This is a very cheap and effective method for the management of Parthenium in uncultivated situation.

Parthenium and Cassia: During a visit to Dharwad in October 1982 and 1983, the author observed in and around the city that *Cassia* spp. was gradually replacing Parthenium plants in small pockets here and there. During 1983, replacement of Parthenium by *Cassia* spp. (Plate 26) had taken place to a larger extent compared to the previous year. Since this finding was of much interest, the observed *Cassia* spp. was got identified as *C. sericea* (*Synonym*: *uniflora*) and a research project proposed was approved by the University of Agricultural Sciences, Bangalore forming a team for studying different aspects of Parthenium control. At this point, it was discovered through literature that a similar observation was already made by Singh (1983) and the potential of biocontrol of Parthenium using *Cassia sericea* (CS) was indicated. It was further discovered that the new waste land weed *Cassia sericea,* which is a native of Latin America, entered India on its own in late seventies (Mahadevappa and Joshi, 1985) and started colonizing here and there in Dharwad, Bijapur and Belgaum districts gradually replacing Parthenium. By 1988, it could spread to other parts of these districts substantially suppressing Parthenium growth. Besides, as observed by the author during 1991, it has started spreading to the neighboring districts like Chitradurga, Shimoga, Davanagere, Haveri, Koppala, Raichur, etc., on its own but substantially supplemented through the efforts of local voluntary workers, organizations and civil societies sowing *Cassia sericea* seeds with the intention of suppressing Parthenium growth.

Research on this subject carried out at the University of Agricultural Sciences, Bangalore, confirmed that the plant *Cassia sericea* could exert allelopathic impact by hindering seed germination and suppressing growth of Parthenium both in green house and natural habitat. It was further revealed that the leachates from leaves, pod walls and seeds were more effective than roots in causing such inhibitions (Mahadevappa and Joshi, 1985; Mahadevappa and Ramaiah, 1988; Mahadevappa and Kulkarni, 1991; Mamatha and Mahadevappa, 1992). The plant leachates of *Cassia sericea* have "Kolines" (compound of plant origin affecting germination and growth of other plant species) which accumulate in the soil consequent to death and withering of this plant and interfere with germination and growth (mechanism is termed as "interference") on only Parthenium.

The visual observation on growth of these two plants (Parthenium and *C. sericea*) over three consecutive seasons (Mahadevappa and Ramaiah, 1988) revealed that there was a gradual decrease in Parthenium population and increase in *C. sericea* population in the three locations experimented. Even the vigour of the Parthenium plants growing amidst *C. sericea* colonies was much low being slender and weak with reduced number of heads. Further, as observed in plants growing in three locations during 1989 rainy season, the Parthenium plants growing amidst *C. sericea* colonies weighed (both fresh and dry) as low as half or even less as compared to those growing in places where there was no influence of *C. sericea* plants. The reduced vigour was also reflected in the number of heads per plant which was about half the normal number. The counts taken on the progressive decline of Parthenium plants in *C. sericea* colonies over the three years revealed some interesting information. In one location where Parthenium plants were pulled out in the first year of study, Parthenium population was reduced in number much faster, showing a ratio of 1:1.5, 1:33 and 1:150. In other two locations the reduction of Parthenium population followed a different trend, being comparatively low. From this, it may be concluded that the *C. sericea* plant is capable of replacing Parthenium gradually over years and the replacement can be faster if Parthenium plants are pulled out in the initial stages of competition.

The scientists of the University of Agricultural Sciences, Bangalore and the voluntary organizations like PROPEL (Programme for Parthenium Elimination, that was active in Bangalore, India) individually and jointly studied this method under field conditions and found promising results in many parts of Bangalore (Hosakote road, Sarakki layout, Jakkasandra, Thindlu village, Hebbal railway station, Domlur, Koramangala, etc.) and also in Tumkur and Mysore districts. *C. sericea*, if established initially by sowing its seeds at the start of rainy season and uprooting and destroying of Parthenium before and after sowing, can suppress Parthenium faster and substantially and perpetuate on its own in subsequent years.

Other botanical agents: Promoting growth of *C. sericea and such other biological agents* in abandoned and non-agricultural lands and in all Parthenium infested orchards would be rewarding and should be preferred for the following reasons. The bioagents have, in their system, allelopathic compounds that are capable of suppressing germination and growth of only Parthenium and not other plants and agricultural crops (Mahadevappa and Joshi, 1985; Joshi and Mahadevappa, 1986; Mahadevappa and Kulkarni, 1991). Besides, these planta have other economic importance such as utility of green plant as soil cover (to prevent erosion), green manure, green leaf vegetable at the young leaf stage, fodder, centering material in civil constructions, dry plant as fuel (Plate 27b), seeds as raw material for gum production etc. Unlike Parthenium, the identified botanical agents do not have any adverse effects on human beings, livestock and plant species other than on Parthenium. As per the work carried out (unpublished) at the University of Agricultural Sciences, Bangalore, *C. sericea* leaves contain 22–28% crude protein and 3.5 to 6.5% fat on dry weight basis. This is comparable to the chemical constituents of *C. tora*, a sister species (Joshep *et al.*, 1988) of *C. sericea*. Therefore, there is a possibility that the latter can also be substituted in pelleting fish feeds by mixing it at 25% without affecting the quality of the feed. This of course needs to be confirmed through experiments.

In case *C. sericea* vegetation or any other agent's growth is not desirable in any particular area, it can be easily eradicated since the seeds are heavy and are disseminated through agents that can be controlled and also the seedlings can be identified in the initial stages and uprooted. However, such a situation may not arise, since these plants are harmless from all angles, unless the land is going to be diverted for a specific purpose. *Mirabilis jalapa* is welcome vegetation since its flowers add to the aesthetic look of the area it covers. *Styloxanthus scabra* or *S. hamata are* a good fodder and a good cover crop in forests and pastures. Like this, all the botanical agents have more uses and are of value to the society in more than one way.

Difficulty in Establishing Botanical Agents

A few botanical agents have been identified to suppress the growth of Parthenium. Many more may be added in due course as the research effort continues to locate the agents suitable for different agro-ecological situations. The example of *C. sericea* comes handy for understanding the problem of establishment. It is growing on its own replacing Parthenium, but the process in the natural course has been very slow both due its bigger seed size and human disturbance uprooting it for various purposes such as green manure, fuel, high way cleaning etc,. It has taken about 10 years to cover sparsely about five districts in North Karnataka during seventies and eighties. Further, it is found difficult to establish in very dry areas and also in elevated places, where the soil moisture is relatively low. If the species is grown in a particular area, as has been explained elsewhere, the leachates accumulated from the plants exert allelopathic impact and suppress the germination and growth of Parthenium. By chance, if the soils where the leachates have accumulated are removed, there will not be any effect of *C. sericea*. Therefore, the Parthenium will start growing and will disappoint the social workers. Further, this beneficial plant can not be established amidst lawns and flower beds since it is an annual plant and dries after the completion of its growth and gives an odd appearance. Yet another point is, it takes time to establish in the initial one or two years which discourages the organizations expecting quick results in controlling Parthenium. Still further delay is expected if the seed dormancy is not broken when sowings are taken up in a fresh area. Like this each botanical agent will pose some difficulties resulting in non acceptance in the initial stages. But once effect is demonstrated in the field, it will catch up and encourage the workers to continue their effort. It is for this reason it is always highlighted that education about the harmful effect of Parthenium and beneficial effect of botanical agents is very crucial in suppressing the growth of Parthenium.

Ramadevi *et al.* (1997) conducted a study on the utilization of allelopathic principles in trees to control Parthenium by spraying the extract of bark and leaf of *Eucalyptus tereticornis* and wood of *Hopea parviflora* plants to study the weedicidal effect. Eucalyptus oil (5–10%) and its high boiling fraction were found very effective in killing growing tips along with flowers within few hours of spraying. The alcoholic extract of leaves also exhibited wilting effect on the plants. The wood extract of *H. parviflora* was not effective. Based on their studies on allelopathic effect of 14 leguminous species suggested that three species namely, *Delonix regia, Prosopis juliflora* and *Saraca indica* have significant allelopathic potential for containing the growth and spread of Parthenium which can be gainfully exploited. The effect of Parthenium leaf extract on seed germination, seed viability of different crop species have been studied by Nagaraj *et al.* (1997), Santha *et al.* (1997) and Banwarilal *et al.* (1997) to find that, extract of Parthenium leaf influences adversely crops like maize, soybean, chickpea, groundnut, rice, green-gram and oat. At certain level of concentration the extracts could produce stimulatory effects on roots, shoot and biomass accumulation. Chakravarthy *et al.* (1997), in a comparative study on the role of *C. sericea* and beetle, reported the suppression of growth in a population of Parthenium plants. The number of Parthenium plants entering reproductive stage was maximum (35.0/1.44 m^2) when defoliated by beetle plots compared to that in *C. sericea* plots. Seedling emergence of Parthenium was maximum in beetle defoliated and weedicide applied plots (1/100 days) compared to 0.3 in *C. sericea* plots. Results obtained in both the years clearly established that *C. sericea* has an efficient mechanism to suppress Parthenium and could supplement the effect of other methods.

b) Insect agents

There are a good number of insect and non-insect pests controlled through biocontrol approach as reported in the literature in the past. Control of *Opuntia* in India through the introduction of cohineal insect from Australia is a successful instance in India. It has been found in Australia, that biological control of the weed is the only cost-effective, environmentally safe and ecologically viable method available. Eight species of exotic insects have been introduced into Australia for the biological control of Parthenium weed, of which at least six species are known to be established. The stem-galling moth *Epiblema strenuana,* the stem-boring weevil *Listronotus setosipennis,* and the leaf-mining moth *Bucculatrix parthenica* are the species that are successfully established in most of the Parthenium infested areas of Queensland, while the leaf-feeding beetle *Zygograma bicolorata* and the seed-feeding weevil, *Smicronyx lutulentus* appear to be established only in the central Queensland region.

Pathogens Specific to Parthenium

Pathogens that are specifically infecting Parthenium **(Tables 16–17)** have been tried for their potential role as biological control agents. Fungal pathogens collected during surveys in the neotropics for coevolved natural enemies of Parthenium, the rust fungi, *Puccinia abrupta* var. *partheniicola* and *P. melampodii* and the white smut, *Entyloma compositarum* are considered as potential biological control agents (Fauzi *et al.*, 1999; Seier *et al.*, 1997; Evans, 1997). Infection with *P. abrupta* var. *partheniicola* hastens leaf senescence and significantly reduces the life cycle and dry weight of plants. The production of mature, seed bearing flowers is also reduced 10 fold on diseased plants, thus *P. abrupta* var. *parthniicola* appears to be a promising bio control agent

Table 16. Pathogens specific to Parthenium

Pathogens	Reference
Fungi	
Puccnia abrupta var. *partheniicola*	Parmelee (1967), Tomely (1990), Parker *et al.* (1994), Evans (1997), Fauzi *et al.* (1999)
Puccinia melampodii	Parmelee (1967), Parker (1990) Evans (1997)
Entyloma compositarum	Ciferii (1963), Evans (1997)
Cercospora parthenii	Chupp (1956)
Odidium parthenii sp. nov	Satyaprasad and Usharani (1981)
Bremia lactucae	Ciferii (1956)
Fusarium pallidoroseum	Kauraw *et al.* (1997)
Bacteria	
Xanthomonas campestris pv. *parthenii*	Ovies and Larinaga (1988)

for Parthnium weed (Parker *et al.*, 1994; Parker, 1990; Evans, 1997; Fauzi *et al.*, 1999). *P. melampodii* has extremely wider host range (Parmelee, 1967) but the provisional results of Seier *et al.* (1997) indicated that there are isolates with restricted host range and acceptably high level of specificity. *Fusarium pallidoroseum* is host specific and found to be a potent and safe biocontrol agent (Farkya *et al.*, 1996; Kauraw *et al.*, 1997; Madhukeshwara *et al.*, 2002). The powdery mildew caused by *Oidium parthenii* (Satyaprasad and Usharani, 1981) and *Bremia lactucae* (Ciferii, 1956), bacterial leaf blight caused by *Xanthamonas campestris* pv. *parthenii* pv. *nov* (Ovies and Larringa, 1988) and leaf spot caused by *Cercospora parthenii* (Chupp, 1956) were also found useful as biocontrol agents.

Past efforts using pathogens as biocontrol agents indicated that Parthenium management through pathogens cannot be as simple as in some other successful cases because of high regeneration capacity, large seed production ability, germination ability throughout the year and extreme adaptability of Parthenium in wide range of ecosystems.

From the above review, it is evident that Parthenium weed is a potential reservoir host for many of the plant pathogens. Recently there was an epidemic of peanut stem necrosis disease (PSND) in groundnut caused by tobacco streak virus (TSV) in Andhra Pradesh. TSV is pollen transmitted in Parthenium and can easily infect groundnut causing severe losses. Inoculum source from Parthenium was found to be one of the major factors in causing epidemic of PSND in groundnut (Prasada Rao *et al.*, 2000; Prasada Rao *et al.*, 2003). Parthenium infected with tomato leaf curl virus was served as potential source of disease inoculums for tomato causing severe infection. Inoculums source from Parthenium served as one of the factors driving tomato leaf curl virus disease epidemics in South India (Govindappa *et al.*, 2004). Eradication of Parthenium in and around the cultivated crops would help in reducing the inoculum source of plant pathogens.

Epidemiological basis of Parthenium allergy

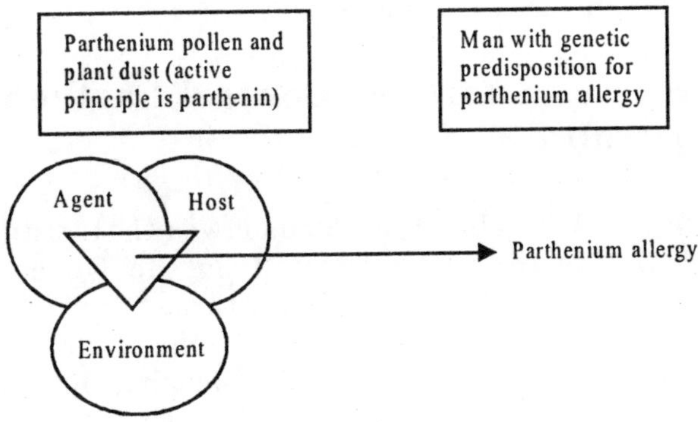

The use of obligate fungal pathogens, such as rusts and smuts, as classical biological control agents of invasive alien weeds has proven to be a highly successful management strategy with an exemplary safety record (Evans *et al.*, 2001; Tomely and Evans, 2004; Barton, 2004). However, there are no short cuts and high scientific standards need to be maintained by biological control practitioners (FAO, 1996), if this exciting and still relatively novel approach to weed control is to gain more general acceptance. India has shown itself to be ready and able to embrace this "technology" in order to tackle its invasive weed problems with the recent release of *P. spegazzinii* against mikania weed. It is confidently predicted that this rust will prove to be the "silver bullet" for mikania control, as previously experienced with rubber-vine rust in Australia, where weed populations over a 40,000 sq km invasive range have been brought under control (Tomely and Evans, 2004) and with white smut of mistflower in Hawaii, New Zealand and South Africa (Evans *et al.*, 2001; Evans, 2002).

Unfortunately, in the case of Parthenium in India, the conclusion is reached that this silver-bullet success will not be repeated and that a guild or suite of natural enemies including arthropods and fungi will have to be introduced, and employed in combination with cultural and chemical control. Management has to be viewed in the long-term, as there is no "quick-fix" solution to this hugely problematic and dangerous weed.

Progress of Efforts to Control Parthenium in Australia

In 1978, in Australia, the Commonwealth Institute of Biological Control was contacted by the Queensland Department of Lands to carry out a three years study on insects attacking *Parthenium hysterophorus* in Mexico. One of the insects from Mexico, a leaf-feeding chrysomelid

Table 17. Fungal pathogens associated with Parthenium

Pathogen	Native range	Exotic range
Obligate fungal pathogens		
Bremia lactucae	Dominican Republic, Mexico	—
Entyloma compositarum (= *E. parthenii*)	Argentina, Dominican Repblic, Mexico	—
Erysiphe cichoracearum	Mexico	India
Oidium parthenii	—	India
Plasmopara halstedii	Dominican Republic, Mexico	—
Puccinia abrupta var.	Argentina, Bolivia, Brazil,	China, Ethiopia,
Partheniicola	Central America, Mexico	India, Kenya, Mauritius, South Africa
P. melampodii	Central America, Mexico	—
Sphaerotheca fuliginea	—	India
Facultative fungal pathogens		
***Alternaria* spp.**	—	India
A. alternate	—	India
A. protenta	Mexico	India

Table 17. *Continued ....*

Pathogen	Native range	Exotic range
A. zinniae	Mexico	India
Cercospora partheniphila	Cuba, Mexico	India
Colletotrichum capsici	—	India
C. gloeosporioides	—	India
Cryptosporiopsis spp.	—	India
Curvularia lunata	—	India
Exserohilum rostratum	—	India
Fusarium spp.	—	India
F. pallidoroseum	—	India
Lasiodiplodia theobromae	—	India
Myrothecium roridum	—	India
Phoma sorghina	—	India
Rhizoctonia solani	—	India
Scleroinia sclerotiorum	—	India
Sclerotium rolfsii	—	India

– *Zygogramma* spp. Nr. Malvae Stal. was found to be host-specific. A second species, the seed-weevil, *Smicrony lutulentus* Dietz was also proved host specific in preliminary testing (McClay, 1980) and in December 1979, it was introduced in Queensland. The biological agents used successfully in Australia are as follows:

- *Zygogramma*, the leaf-defoliating beetle
- *Listronotus*, a stem-boring weevil
- *Simcronyx*, a seed-feeding weevil
- *Epiblema*, a stem-galling moth
- *Bucculatrix*, a leaf-mining moth
- *Carmenta*, a root-boring moth
- *Puccinia abrupta* var. *partheniicola*, the winter rust and Puccinia melampodii, the summer rust.

With the adoption of this technology, the weed could be brought under check. The studies on biology and ecology involved the characterization of ecotypes using genetic finger printing techniques, investigating the role of allelopathic, seed banks and phenological attributes in the weeds persistence mechanism(s). Process-based simulation models and geographical information systems are used to monitor and predict future spread. Biological control is developed and enhanced through the use of plant feeding insects and pathogens. Extension is carried out through the already existing networks of the CTPM and the central Queensland Parthenium Action Group (PAG).

The new initiative program on the biological control of Parthenium is currently mapping the distribution of various biocontrol agents in Queensland, and assessing the role of already established biocontrol agents in controlling the weed populations. The stem-boring weevil *Conotrachelus* spp. and the stem-boring moth *Platphalonidia mystica* both from Argentina are being released in the central Queensland region. In addition, two new root-feeders, *Carmenta ithecae* and *Thecesternus hirsutus,* which are being imported from Mexico will be host-tested. The newly formed Parthenium Action Group is involved in the distribution and field evaluation of already established biocontrol agents in the central and northern Queensland regions (Dhileepan *et.al.*, 1996). One of the fungal pathogens underwent host-range testing in the United Kingdom and was released in Australia (McClay *et al.*, 1995).

Insects feeding on Parthenium: Many insects are observed to feed on Parthenium **(Table 18).** Some of them as reported from different places are mentioned below:

The nymphs and adults of mealy bug, *Ferrisia virgata* Cockrell were reported to feed on roots of Parthenium in Mysore city. The young plants were prone to heavy attack by the bug and the plants died completely due to wilting (Char *et al.*, 1975). Krishna Murthy *et al.* (1976b) have also reported similar observation in Hassan. Even the vegetable mite (*Tetranychus cucubitae* Rahman and Sapona) has been observed to infest this weed on the farm of Agricultural College, Dharwad (Puttaswamy *et al.*, 1976), since these are polyphagous, they cannot be used as biological control agents.

In 1983, the Indian Institute of Horticultural Research introduced the leaf-feeding beetle, *Zygogramma biocolorata* (*Coleoptera*: *Chrysomelidae*) from Mexico. Detailed host specificity tests carried out under quarantine conditions confirmed that the insect does not harm cultivated crops in the country. Earlier studies in Australia, where Parthenium was posing serious threat and also in Mexico have established that this beetle is incapable of multiplying on any plant except Parthenium. Based on these results, field release permits were obtained from the plant protection advisor in 1984 (Jayanth and Nagarakatti, 1987).

Table 18. Status of *Zygogramma* spp.

Name of the species	Status
Zygogramma bicolorata (Pallister)	– Feeds of *Parthenium hysterophorus* (McClay, 1980) – Introduced in Australia and India (Jayanth, 1987; McFadyen, 1992)
Z. disrupta (Rogers)	– Introduced in USSR against rag weed *Ambrosia* sp. (Kovalev and Medvedev, 1983) – Feeds on rag weed in USA (Piper, 1978)
Z. exclamationis (Fabricius)	– Feeds on wild and cultivated sunflower plants in USA (Regers and Thompson, 1980)
Z. suturalis (Fabricius)	– Introduced against rag weed *Ambrosia* spp. In USSR (Kovalev and Medvedev, 1983) – Feeds on rag weed in USA (Piper, 1975)
Z. tortuosa	– Introduced against rag weed. *Amvrosia* spp. in USSR (Kovalev and Medvedev, 1983) – Feeds on sunflower under laboratory conditions culture destroyed in USSR for fear of attacking sunflower – In california tested on sunflower in laboratory. – Larvae failed to complete development died in first or second instar. In field, adults left the plant and died after 3 days (Goeden and Ricker, 1979).
Z. conjuncta (Rogers)	– Described under genus Zegospila. – *Z. conjuncta* feeds on poverty weed *Luva axilaris* Pursh (Best, 1963)

Source: Singh (1997)

The adult *Z. biocolorata* measures about 6mm in length (Plates 28a), lays eggs singly or in small groups on the under-surface of leaves of Parthenium (Plate 28b). The eggs hatch in 4–6 days and the larvae (Plate 28c) feed voraciously on Parthenium, first attacking terminal and axillary buds and later the leaf blades. The young larvae check plant growth and flower production due to severe defoliation. Full grown larvae desert the plant, drop off and pupate in the soil (Plate 28d). The insect completes its developmental period in 25–30 days. The adult beetles lay an average of 2000 eggs during their life span of three or four months. *Z. biocolorata* remains active in the field during rainy season from June to October. The beetle can tolerate dry spells in between, although their multiplication rates may be affected. The weed killer is capable of overcoming winter and summer months by getting into the soil and diapausing there for 6–8 months. The diapaused beetles come out with the onset of monsoon rains in June. Thus there will be no need for reintroduction of the beetles in areas where they are already present (Jayanth and Geetha Bali, 1993). Recently it is reported from the Indian Agricultural Research Institute that there are biotypes which are capable of surviving and feeding even during summer months around Delhi.

Field releases of *Z. biocolorata* were initiated in Sultan Palya in Bangalore in June, 1984. Although the insect established readily, no significant impact could be obtained during the first four years. However, since 1988 there has been a dramatic increase in field population of the beetles resulting in large scale defoliation of the weed in many parts of Bangalore city (Plate 29). By 1991, it could spread on its own and through releases by voluntary organizations and Research Institutes to nearby place which are about 40–50 km away from Bangalore city, causing significant damage to Parthenium populations by fast defoliation. Subsequent releases across the country have demonstrated that it can adapt to many situations and supplement towards control of Parthenium. The author has observed it in Himachal Pradesh, Haryana, Kulu, Andhra Pradesh and Maharashtra.

Effectiveness of releasing beetle (*Zygogramma biocolorata*)

- The insect is active during the rainy season only for about 5months in a year, whereas the Parthenium plant can germinate and grow round the year (Jayanth and Geetha Bali, 1993).
- Even unseasonal rains do not cause termination of diapauses (Jayanth and Geetha Bali, 1993). Hence, it can not help in controlling the winter and summer growth of Parthenium.
- The pattern of feeding Parthenium plants is very erratic always leaving a portion of the plant population untouched. This gives room for seed multiplication and spread.
- The Parthenium plants covered by dust and the rosettes (formed because of cutting the immature plants and due to drought during the growth stage) are not eaten by the beetle and such escapes produce seeds when the conditions become favorable.
- The biocontrol method can work more satisfactory in cases where the host organism has a limited adaptability (like Opuntia among plants which grow only in specific ecosystem, crops which are grown only during a particular season, insect is active only in specific situation). It is feared that the beetle expands its host range and takes to other economically important plants in course of time. Chakravarthy and Bhat (1994) reported that the beetle was feeding on sunflower in Chintamani of Kolar district. Sridhar (1991) published a detailed

report on the feeding of Mexican beetles on sunflower. Such fears were cleared by an expert committee constituted by Government of India during eighties.

Considering the severity of problem, the Indian Council of Agricultural Research (ICAR), New Delhi constituted a fact finding committee on Parthenium, under the chairmanship of Dr. V.M. Bhan (Director, NRC on Weed Control, Jabalpur). The committee considered the various issues related to the controversy raised regarding the feeding of *Z. bocolorata* on sunflower (Anon., 1993).

- The committee opined that *Z. biocolorata* feeds only on the border rows of sunflower due to deposition of Parthenium pollen and therefore, is unlikely to cause economic damage. The benefits derived due to the suppression of Parthenium by the beetle outweighed the insignificant damage on sunflower. Therefore, the embargo imposed on its multiplication and release may be removed. Accordingly the Chief Plant Protection Advisor, Govt. of India has removed the embargo and has directed central IPM centers for mass multiplication and release of the beetle.
- The beetles are not active fliers; as such the rate of spread in space and time is rather slow as compared to the spread of the host.

Parthenium can adapt to a wide variety of situations. Therefore it can not be controlled by any one or two agents. But the successful release of many insects in Australia to feed on seed, stem, flowers, roots etc and successful suppression in most places is a classical example that can be tried else-where also.

Controversy over feeding of beetle, *Zygogramma bicolorata* on sunflower: Seven years after first release of this beetle, it was found feeding on an important oilseed crop, sunflower (Sridhar, 1991) followed by other reports (Narendra, 1993; Kumar, 1992). In view of this controversy, two independent Fact Finding Committees (FFC) were set up, one by the Government of Karnataka in February 1992 and other by the Indian Council of Agricultural Research (ICAR) in November, 1992. In its first meeting in 1992, FFC (ICAR) dissolved.

On the basis of developmental studies conducted at various centers co-ordinated by FFC, it was unequivocally proved that *Z. bicolorata* is a safe bioagent against Parthenium. Little feeding by the 'O' day beetles and freshly hatched grubs on sunflower near the border rows of heavily infested Parthenium was attributed to falling of Parthenium pollen on sunflower which attracted beetles to feed. Continuous feeding on sunflower caused degeneration of the ovary (Jayanth *et al.,* 1997; Sushikumar and Bhan, 1998).

Studies revealed that in spite of development of *Z. bicolorata* on sunflower for several generations in the laboratories, chances of beetle to become a potential pest was remote as the survival rate of grubs and beetles developed on sunflower was very less.

Field experiments conducted at Vindhyanagar in a Madhya Pradesh clearly demonstrated that beetles did not feed on sunflower grown amidst heavily infested Parthenium colonies.

The NRC Jabalpur is multiplying this insect and distributing throughout 22 centres spread across the country. The centre has also distributed to about 150 out of 400 krishi vigyan kendra (KVKs). The centre supplies beetle with required technical information to any place in the country on request.

End of the controversy on feeding potential of Mexican beetle on sunflower: The fact finding committee in its seventh and last meeting held at

Bangalore in May 1998, unequivocally recommended to Government of India that *Z. bicolorata* fed only on the border rows of sunflower under specific situations and did not cause economic damage. Embargo imposed on its multiplication and release was lifted and so also the ban imposed on the Mexican beetle. Now Mexican beetle can be multiplied and released anywhere in India for Parthenium suppression. The release of this beetle is supplementing other methods of control in different parts of India.

C) Virus and MLOs

Some scientists have held the view that there may be few viruses that can be used in biocontrol of Parthenium. Parthenium phyllody disease is very common in Parthenium weed in India. The incidence varies from 10 (February–June) to 100% (August–December). The leafhopper (*Orosius albicinctus*) population in the field is positively correlated with the incidence of this disease. The phyllody disease of Parthenium is transmitted by *O. albicinctus* and the active transmission is found to be 55%. The minimum acquisition access period is found to be 20 min and the inoculation access period is 15 min. Incubation period in the vector varies from 15 to 20 days. The pathogen persists throughout the life of the *O. albicinctus* (Mathur and Muniyappa, 1989). Parthenium phyllody was transmitted by *O. albicinctus* to aster, cowpea, blackgram, greengram, horsegram, limabean, redgram, sannhemp, fieldbean, soybean and wingedbean (Mathur and Muniyappa, 1989). Mycoplasma diseases occurring naturally on these crop plants were transmitted to Parthenium by *O. albicinctus*. Scientists researching on this area feel that there is no hope of using this as a biocontrol method in the near future, but their efforts are on with a hope of positive results in future.

Limitations of Different Methods

Some scientists have advocated two more methods *viz.* cultural practices and growing competitive crops. These are impracticable for the following reasons. Operations such as ploughing, passing of cultivator/harrow and intercultural operations cannot be practiced in the lands with heavy Parthenium infestations which are usually waste lands or no man's lands. Also such operation incurs heavy recurring expenditure. The only situation where cultural practice can work is orchards like guava, coconut, mango etc. where, usually, such operations are done traditionally and all weeds including Parthenium are controlled in a routine way with scheduled seasonal operations.

Growing competitive crops is also not practicable in agricultural lands, because, Parthenium is not a big problem in cultivated lands where regular weeding operation is done. Further, cropping pattern has to be decided based on the suitability. Cropping sequence, crop compatibility and marketability and changing the crop for the sake of Parthenium control is not going to become practicable. One possibility of using a plant enemy in intensively cultivated land to control Parthenium is only growing of *Styloxanthus scabra* and *Tephrosea* spp. in lands exclusively devoted to forage production. Some information of relevance on this aspect can also be found in the section relating to biological control of Parthenium through plants.

d) Integrated Parthenium weed management (IPWM)

This management tactic may be the most effective method of Parthenium control. Under this method, all the available techniques have to be integrated as per the location requirement and locally available resources.

However, as Parthenium is more a weed of wasteland, fallow land and vacant land, bio-intensive integrated management will be the most suitable option. Mahadevappa (1997) and Sushilkumar and Sarawsat (2001) opined that integrated management is the only solution to overcome the menace of Parthenium. They proposed a scheme of integrated management which is presented below (National Research Centre for Weeds-Annual Reports). This method is dealt with in detail in Chapter 8, as this has been found practicable in many situations, and also technology needs to be refined for a given situation.

The adaptive research carried out jointly by the University of Agricultural Sciences, Bangalore and Programme for Parthenium Elimination (PROPEL), Bangalore, in and around Bangalore city for 5 years commencing from 1986, involved employment of different control methods individually and in combination over large areas. The studies clearly demonstrated that only an integrated approach now termed "Integrated Parthenium Weed Management (IPWM)" involving the various methods suggested in the past could be effective in controlling Parthenium. If a concerted effort is made to adopt IPWM, the results can be seen in the second year, and, by the end of third year the Parthenium will come down to a negligible level. The IPWM technology envisages five steps *viz.*, (i) Maintenance of natural biodiversity *i.e.*, not disturbing the existing flora to the extent possible. (ii) In places where cleaning and exposing of soil is unavoidable, planting of proven botanical agents (like *C. sericea* or such other plant species) may be taken up at the start of rainy season. The growth of such plants can insulate opened up soils against invasion by Parthenium. In already infested areas, planting of botanical agents may be taken up at the start of rainy season and Parthenium plants which may grow along with the botanical agents are to be removed preferably before flowering during the first one or two years of starting adoption of this technology, so that the antagonistic plants establish well. Afterwards, there will be no need to plant the intended botanical agents again as it will perpetuate on its own. (iii) To watch for the commencement of rains and build up of beetles and when the beetles become available in large numbers, they have to be collected and released in other Parthenium invaded areas. (iv) In situations where none of the above methods can be adopted for aesthetic or other reasons (as in case of gardens, flowers beds, lawns, intensively cultivated agricultural fields), manual removal has to be resorted to by identifying the persons who are not allergic to this plant. (v) In situations where none of the above methods can be adopted like vast stretches of already Parthenium invaded plots and where immediate relief is needed, it may have to be controlled through chemical sprays with the knowledge that it gives only temporary suppression and does not ensure permanent control. Further, as chemical method causes environment pollution, it is recommended only as a last resort under unavoidable situations, when emergent situations arise for cleaning Parthenium populations. The last two methods can also be integrated initially with biological methods which essentially is IPWM, in order to achieve desired results faster.

Operational Tips for IPWM

1. Mark the boundary of the area to be treated.
2. Uproot Parthenium plants without disturbing the naturally existing vegetation.
3. In open soils (where there is no vegetation) open shallow furrows of about 2.5 cm deep and sprinkle seeds of botanical agents (any other proven botanical agents) at fixed seed rate. If it is *Cassia* species, furrows opened at 22.5 cm apart are optimum for the purpose of sowing establishing to be effective with in a short period.

4. When the seeds germinate after rains, encourage the growth of botanical agents by protecting from trampling or any other kind of physical damage and uprooting only Parthenium plant before flowering.
5. Pay special attention to protect and promote the growth of the following plant species which exist in the waste lands in natural course: *C. sericea* (*syn*: *C. uniflora*), *C. tora*, *C. auriculata*, *Croton bonplandiamum* (*syn*: *Croton sparsiflorus*), *Amaranthus spinosus*, *Tephrosia purpurea*, *Hyptis suaveolensi*, *Sida spinosa*, *Mirabilis japala* and such other botanical agents identified from time to time.
6. Uproot Parthenium plants at intervals so as to enable botanical agents to put up full growth.
7. Once these are accomplished, they perpetuate on their own and there will be no need to plant them again unless the soil is scraped to take out self sown seeds of biological agents.
8. The sowing of seeds can be taken up just before the onset of rains or after the first rains.
9. It is to be ensured that seeds of biological agents collected and stored are scarified for breaking seed dormancy before sowing, as these plants are grown wild hence have seed dormancy of varying period. This is needed in the initial plantings only. In subsequent years, the seed dormancy gets broken in the natural course and hence no effort is needed towards breaking dormancy.

e) *Success stories of Parthenium control by bioagents in Karnataka*

With a view to popularize a low cost and effective technology for controlling Parthenium through out the state, the University of Agricultural Sciences, Dharwad, collected 70 kg of *C. sericea* seeds, scarified them and supplied through the Joint Directors of the Department of Agriculture of different districts and the Karnataka State Seeds Corporation during 1998. Later, observations on the Parthenium populations were recorded by the scientists of University of Agricultural Sciences through a survey of the districts.

In a farmer's field at Kolihalli industrial area and another field in Hirehalli watershed area of Tumkur district, it was observed that about 80% of Parthenium was controlled. The density of Parthenium plants was found to be 12 plants/m^2 in the absence of *C. sericea*. Line sowing of *C. sericea* seeds produced better results than broad casting in Chinkurali watershed area. Excellent germination of *C. sericea* was noticed in Doddayaraganhal village of Arasikere taluk. *C. sericea* in this area was sown on a stretch of 2 km on either side of Mysore road at a very low cost of Rs. 50. In a heavily Parthenium infested coconut orchard at Kadur, considerable control of Parthenium was reported over a period of eight years. Sowing of *C. sericea*, which later gave the density of 60 plants/m^2 effectively controlled Parthenium population in a heavily Parthenium infested cultivable land at Musukinakatte village in Chikmagalur district.

Other botanical agents such as *Cassia tora*, *Styloxanthus scabra, Sida spinosa, Hyptis suaveolense, Tephrosia purpurea* (Plate 30), *Amaranthus spinosus, Cassia auriculata* and *Croton sparciflorus* (Plate 31) were also noticed in abundance in Chitradurga, Tumkur, Hassan, Chikmagalur and also in Belgaum districts. After a reconnaissance of botanical agents in Karnataka, Kolar district has been identified as the best area for collection of botanical agents other than *C. sericea*.

In India, the effect of release of beetle is seen to be varying in different situations. There is no definite survey undertaken to quantify the extent of control in different parts. The author's own observation across the country during his visits in different years has confirmed that during the peak season of Parthenium growth the infectivity is more resulting in defoliation ranging from 100 to 10%. One special observation is that in many places, there was significant difference between two sides of the road. It is to be ascertained whether it is due to wind and any other factor. One important observation is the spread of insect and consequent damage is increasing year by year. No effort is made to record occurrence of insect and damage caused on state or region wise. One observation made during 2005 in Karnataka revealed that damage to Parthenium plants by *Z. bicolorata* was more than 50% in Mandya district, while in Hassan district it ranged from 60–90%.

PLATES 1-7

Pl. 1(a). Parthenium plant

Pl. 1(a). Parthenium plant

Pl. 2. Parthenium competing with agricultural crops

Pl. 3. Parthenium spread in city areas

Pl. 4. Parthenium spread in human habitats

Pl. 5. Parthenium growth along the roads

Pl. 6. Parthenium growing abundantly in cultivated fallows

Pl. 7. Luxuriant growth of Parthenium in young orchards

PLATES 8 -13

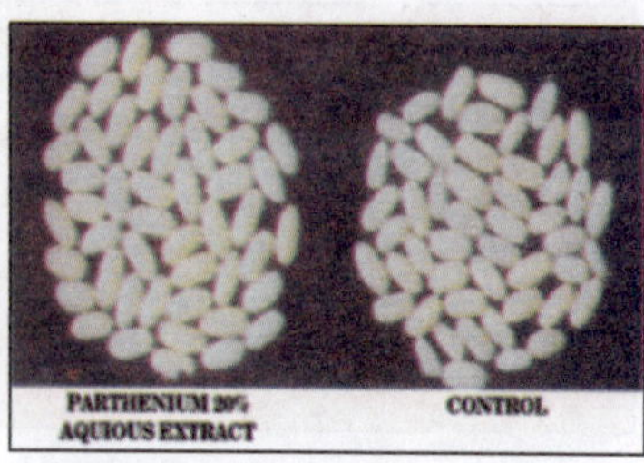

Pl. 8. Beneficial effects of aqueous extract of Parthenium on mulberry silk worm cocoons

Pl. 9a. Parthenium exposure to street children and homeless dwellers

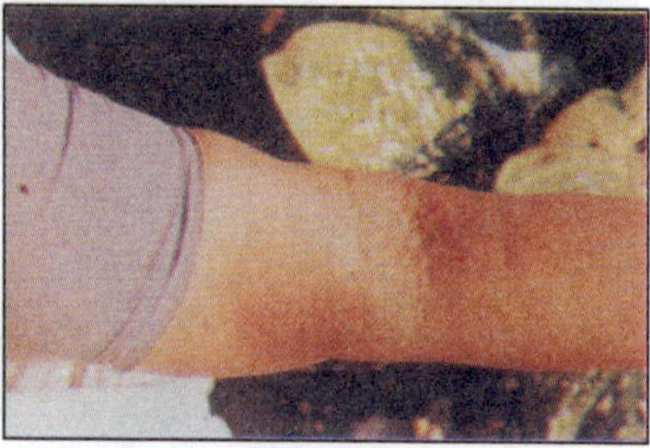

Pl. 9b. Allergic reactions on the arm of a woman

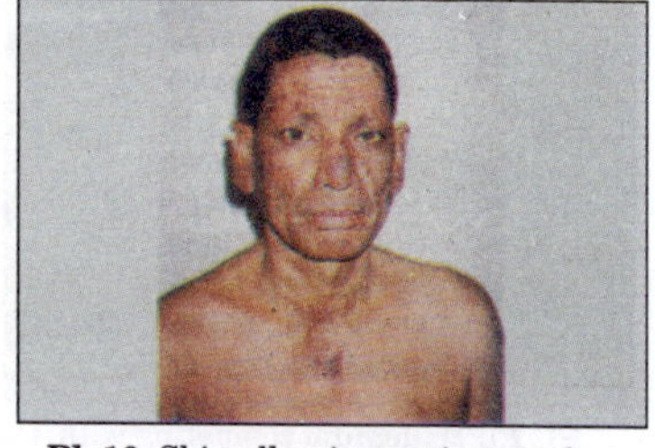

Pl. 10. Skin allergic reactions on face due to Parthenium

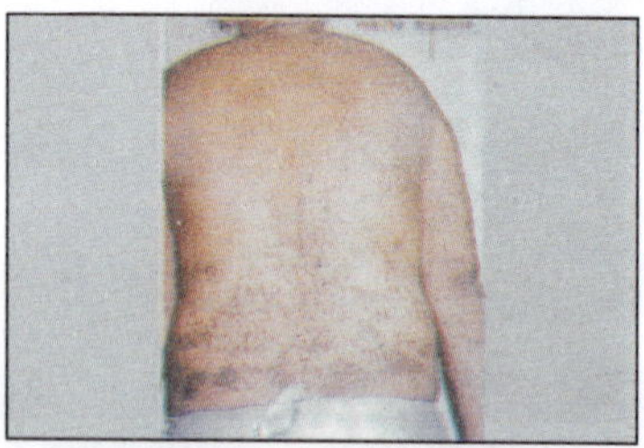

Pl. 11. Skin allergic reactions on the back of human body

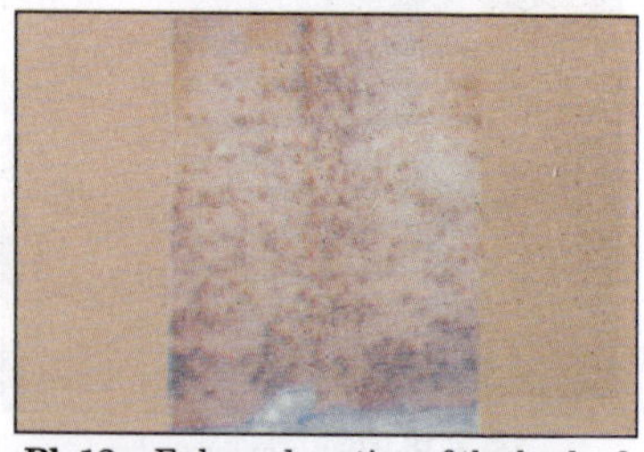

Pl. 12a. Enlarged portion of the back of human body with allergic symptoms

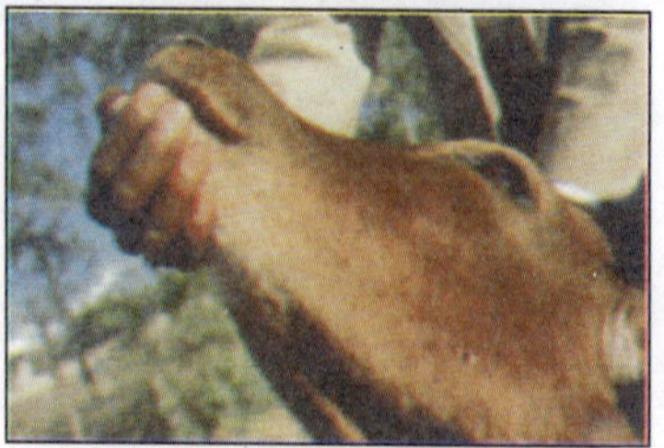

Pl. 12b. skin allergy developed on animals

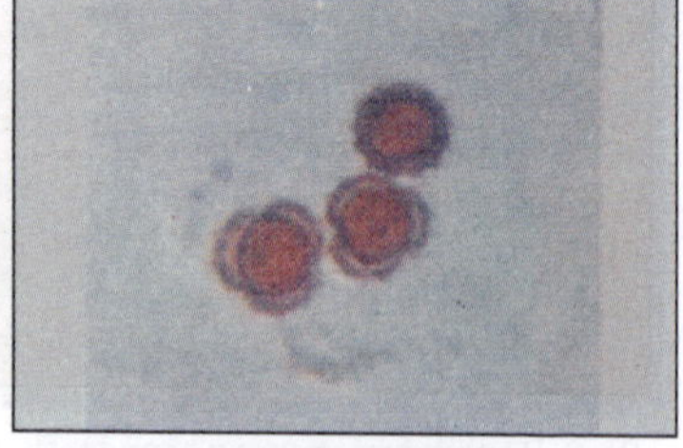

Pl. 13. Pollen grains of Parthenium

PLATES 14-21

Pl. 14. *Cassia sericea* plant

Pl. 15. *Tephrosia purpurea* plant

Pl. 16. *Styloxanthes scabra*

Pl. 17. *Croton sparciflorus*

Pl. 18. *Hyptis suaveolens*

Pl. 19. *Amaranthus spinosus*

Pl. 20. *Sida spinosa*

Pl. 21. *Cassia occidentalis*

PLATES 22-28

Pl. 22. *Cassia tora*

Pl. 23. *Cassia auriculata*

Pl. 24. *Tagetus erecta*

Pl. 25. *Mirabilis jalapa*

Pl. 26. Suppression of Parthenium by *Cassia* along the road side

Pl. 27(a). Grown up *Cassia* being cut for fuel wood purpose

Pl. 27(b). *Kochea indica* — Fuel for poor in winter

Pl. 28(a). Adult of *Zygogramma bicolorata* beetle

PLATES 28-33

Pl. 28(b). Eggs of *Zygogramma bicolorata* beetle

Pl. 28(c). Grub of *Zygogramma bicolorata* beetle

Pl. 28(d). Pupa of *Zygogramma bicolorata* beetle

Pl. 29. Large scale defoliation of Parthenium by the Mexican beetles

Pl. 30. Suppression of Parthenium around the tank embankment by *Tephrosia* spp.

Pl. 31. Replacement of Parthenium by *Croton sparciflorus*

Pl. 32. Suppression of Parthenium by *Amaranthus spinosus*

Pl. 33(a). Cases of Misuse; **(b).** Cases of Misuse

Seven

WEED CONTROL ACT

Considering the hazards of Parthenium, Government of Karnataka took the lead in covering this weed under the "Karnataka Agricultural Pests and Diseases" Act., 1969.

Several questions had been posed to the Hon'ble Minister for Agriculture and Irrigation in Rajya Sabha and in the Parliament, New Delhi (on 21st January, 1976) regarding control of this weed. Likewise, the Members of Legislative houses of Maharashtra and Karnataka enquired the Hon'ble Ministers for Agriculture regarding its eradication. All these indicate the awareness and the concern on the part of the members, public and the Government in the matter. With regard to its implementation, notices were issued twice in Bangalore, during the eighties by the City Municipal Corporation, but there was no follow up action to understand the response or to strictly enforce the Law.

A national level task force needs to be constituted and adequately supported to take up measures to check the invasion and further spread of Parthenium weed, as is done in New South Wales (NSW) and a management plan document needs to be developed scientifically for quick results. The document should specify action to be taken when dealing with new outbreaks, placing responsibilities for monitoring and control and provide guidelines for funding. The group should ensure that the weed receives adequate publicity to maintain high profile so that early identification is made. Mr. Kelly attributes the success of Parthenium control in NSW to awareness programmes and the commitment of farmers and Government bodies. Over $ 100,000 were spent in NSW for eradicating the weed. Similarly, the State Government and local authorities in Queensland spent about $ one million in 1989 to control Parthenium weed. The efforts made in Australia by preparing and distributing educative materials with information about the mode of entry of the seeds of Parthenium to newer areas, cautions to be exercised at all stages to prevent dispersal of Parthenium seed and multipronged approach to employ biocontrol methods in an integrated way is given below:

Caution to the public by distributing millions of brochures to reach every citizen in the country. An example of aggressive campaign to educate the public against Parthenium is given below:

Be aware:

- when buying crop seed or stock feed
- when selling grain, seed or hay
- of the origin of stock
- of the origin of machinery and vehicles
- of isolated outbreaks
- That drought heightens the movement of livestock and fodder, and consequently Parthenium seed.

Further, regarding release of different biocontrol agents the role of each agent was made known to the public by distributing brochures with the following information:

For effective suppression of Parthenium, release

- *Zygogramma bicolorata* the leaf-defoliating beetle
- *Listronotus seto-sepennis* a stem-boring weevil
- *Simcronyx lutuentus* a seed-feeding weevil
- *Epiblema strenuana* a stem-galling moth
- *Bucculatrix parthenica* a leaf-mining moth
- *Carmenta*, a root-boring moth
- *Puccinia abrupta* var. *partheniicola*, the winter rust and *Puccinia melampodii*, the summer rust.

Thus the public could participate with confidence and help in containing Parthenium successfully. Other countries take this model for efficient transfer of Parthenium control technology.

Eight

COORDINATION OF CONTROL PROGRAMME

Though, there are efficient control methods now available to suppress the growth of Parthenium and also legal acts in some countries providing for strict measures to enforce the same, there are some practical difficulties in the way of implementation of the same. The primary factor is that the Parthenium infests public land or no-man's land, not cared for in a routine way to keep it clean in the manner the owners of agricultural land do in their own cultivated lands or private land owners do in their activity spots. The control programme should be a concerted and properly organized, involving government sponsored institutions, public and voluntary organizations as also other civil societies.

There is also a need for seminars, training programmes at state and national levels to develop educative material to be used by various media as has been done in Australia.

Setting up of a Parthenium Eradication Cell (PEC) by the concerned government with a few squads for specific purposes will go a long way to achieve good results. The survey squad of the cell should make a thorough survey, spot on the map and keep record on its presence, intensity, spread, etc. The education squad of the cell should educate all the concerned (public, private and voluntary organizations) on possible eradication by highlighting the success stories. The enforcement and control squad should effectively enforce the act for eradicating the weed. The PEC shall have the responsibility to coordinate the control programme, so that Parthenium does not make re-entry into the once controlled area. To work at the national level to educate on the adoption of the latest methods available in controlling this weed in regional languages, within a time frame set for the purpose so that the pernicious weed is suppressed and kept at a safe level. This is possible, with technology available today for adoption under certain situations, in about 3–4 years period, if only a concerted effort is made and measures taken up on a war-footing basis. The basic intention must be to inculcate a sense of participation in the general public and create awareness about the problem of Parthenium and the effective control measures now available.

The "World Parthenium Study Group" which was set up with Dr. Joha C. Mitchell of Canada in seventies with the main objective to exchange the scientific ideas needs to be reactivated and accelerated for its activity (Krishna Murthy *et al.,* 1977).

In conclusion, there is need for people and residents' welfare associations to put their heads together and declare a war on Parthenium so that further growth in city areas could be checked.

Success Stories in Parthenium Management

Efforts to control Parthenium growth in and around Bangalore city started by programme for Parthenium elimination (PROPEL) in 1986-88. Initially, a few kilograms of *C. sericea* (CS) seeds were procured from Dharwad and sown in vacant sites and waste lands in some parts of Bangalore city, through voluntary organizations and individuals. The University of Agricultural Sciences supported with technology back up. In the subsequent years, the quantity of *C. sericea* seed obtained increased to quintals and sowing was done in larger areas excluding vacant sites. Because, the sites were either used for ongoing constructions or dumped with debris, thereby coming in the way of establishment of *C. sericea*. From the year 1988–89, release of beetle, *Zygogramma bicolorata* was integrated with the sowing of *C. sericea* in some places like Hebbal railway station, Amruthahalli cross road (Bellary road), Jakkasandra and Sarakki layout, and the method has proved successful. In some areas like Indira Nagar, the results have been discouraging. In highways like Bellary road, Tumkur road, Bannerghatta road, the methods proved to be partially successful. For the first time, an intensive sowing of *C. sericea* was taken up during 1990, from May to July on most of the highways. A survey conducted in the mid-August during 1990 and 1991 on the establishment of *C. sericea*, beetle activity and the stand of Parthenium revealed the following:

1. *C. sericea* established well by 1991 in 90% of the places, where it had shown partial establishment in the previous year, *viz*., Hebbal railway station, Amruthahalli cross road and parts of Bellary road. The growth of *C. sericea* during 1990 was not as vigorous as in the previous years because of a prolonged drought.

2. From among the areas sown during 1990, very early sown (May) places showed good establishment of *C. sericea* although the growth in terms of plant height was normal in some places. In places where intermittent rains were received, establishment and growth of *C. sericea* were very satisfactory, both during 1990 and 1991. This plant species could spread on its own and extend its area in 1:5 proportion.

3. In places where the soil was very porous and rains were scanty, both germination and establishment were far from satisfactory during 1990. But, in patches where there were ravines, open drains and depressions, the establishment was better and these plants multiplied in 1:10 proportion by 1991.

4. Sowing was also done in some places where the soil was very compact and trampled daily by passers-by. Although the *C. sericea* germinated in these places, establishment was poor for want of proper soil conditions; even Parthenium did not establish itself normally in such situations. Hence it was felt unnecessary to take up any operation in future.

The above areas were observed closely to find out the reasons for good, moderate and poor germination and establishment of *C. sericea* in different locations.

1. Successful establishment of *C. sericea* in some places mentioned above could be attributed to early sowings, followed by enough rains for good germination, growth and root formation before the commencement of long drought spell. To some extent, the moisture retention capacities of the soil and low land situations had also contributed to its successful growth.

2. Non-establishment of *C. sericea* in some places could be attributed to late sowings (July), which might have led to a situation wherein seeds germinated up to two leaf stage (with one or two rains) followed by long drought spell (4 to 5 weeks) leading to death of seedlings.

3. Partial establishment of *C. sericea* in some places was either due to some intermittent rains during the drought spell or due to high moisture content in soil itself due to high water holding

capacity or seepage from the neighborhood drains, ponds etc.

4. In places, where sowing was not followed by early rains, the entire process of germination and growth of *C. sericea* had commenced only after the revival of rains in the month of August (second rains) and germination and growth was very satisfactory. The *C. sericea* plants growing from second rains, however, could not put up normal growth and produce as many seeds as growing from the beginning of the season (June–July). Still, there were enough pods (fruit formation) to multiply the plant population in a small proportion for the next year.

In places where the beetle, *Z. bicolorata* had been released, defoliation had taken place and the results were clearly visible even to a layman. The insects however did not move to western side from places where they were released, but they moved to some extent to the eastern side. None to 50 number of beetles were counted per plant, wherever beetle activity was observed. Considerable reduction in Parthenium seed formation could be seen because of the beetle activity. But this was limited only to the rainy season *i.e.*, from July to November. Parthenium started reappearing wherever it was controlled by insects but it was not so wherever it was controlled by *C. sericea* and other botanical agents.

An Analysis of the Situation

The long drought spell during 1990 was a big handicap for the establishment of *C. sericea* and also for the multiplication of the beetle. Despite this unusually prolonged drought, successful establishment of *C. sericea* and spread of beetle in some places indicated the potentiality of the method in controlling Parthenium growth. The prolonged drought intervals are not very common though intervals are not uncommon. The effect of partial and sparse establishment of *C. sericea* in some places may lead to some disappointment for not showing the expected results. Because *C. sericea* leachates that was expected to accumulate in soil to suppress germination and growth of Parthenium was inadequate. Wherever the *C. sericea* had germinated, but remained without further growth due to stunting in the initial stage could not contribute enough leachates to suppress Parthenium.

Follow up Action

In some areas, where the seedlings dry up due to drought, the entire process of sowing needs to be repeated in subsequent years. In areas, where *C. sericea* has established partially, supplementary sowings in the coming years will help quick establishment of *C. sericea*. In areas where *C. sericea* has established with normal growth, there is no need for fresh sowings.

Some corrections needed in the next sowing: Highly compact and over trampled soils like playground should not be selected for sowing *C. sericea* because neither Parthenium nor *C. sericea* can grow there. The areas likely to be dug for construction of roads, buildings, etc., may also be avoided.

1. Spacing between rows which is 3–8" should be reduced to 2" while sowing *C. sericea*.
2. Thick sowing should be avoided. Only 1/4th of the present quantity of seeds per unit area should be sown *i.e.*, 6–8 kg seeds are enough for planting one acre. This will help to cover more area with the same quantity of seed, besides avoiding inter-plant competition and aiding better growth of the young plants.

3. As far as possible, early sowings and even dry sowings may be taken up (April to June) so that *C. sericea* plants could grow strong enough to withstand any drought intervals.
4. For better results, the Parthenium plants are to be uprooted before sowing *C. sericea*.

The integrated approach being adopted at present has high potential to control Parthenium in the targeted areas, as evidenced by success seen to the extent of 60 to 80% in various districts especially in northern Karnataka and southern Maharashtra.

Nine

INTERNATIONAL CONFERENCES ON PARTHENIUM MANAGEMENT

Two international conferences have been held on Parthenium management so far, the first one during 1997 and second one during 2005. The recommendations of the two conferences are as follows:

First International Conference

The First International Conference on Parthenium Management was held in the main campus of the University of Agricultural Sciences, Dharwad from October 6-8, 1997 (Plate 30). The conference was organized by the University of Agricultural Sciences, Dharwad in collaboration with Indian Council of Agricultural Research, New Delhi, Council of Scientific and Industrial Research, New Delhi and Ministry of Environment and Ecology, Government of Karnataka, Bangalore. The conference was sponsored by the National Bank for Agriculture and Rural Development (NABARD), Mumbai. Eminent scientists, officers of the development departments, progressive farmers and representatives of non-governmental organizations from Australia, United Kingdom and India numbering 120 participated in the three day's conference.

The conference was organized in five technical sessions, wherein 19 oral papers and 54 poster papers covering the following themes were presented.

- Global view of Parthenium and its management
- Advances in Parthenium management
- Utility values of Parthenium
- Future strategies and course of action

The recommendations of the conference are:

1. That an International Parthenium Network (IPAN) be established and a steering committee be elected, comprising representatives from the Governments in affected countries as well as from NGOs and environmental and health groups.
2. That a database of scientists working on Parthenium weed worldwide be setup.
3. That a database of all published and unpublished work on Parthenium weed be created.
4. That an awareness campaign, aimed at both potential sponsors as well as the public in general, be initiated to stress the menace posed by Parthenium weed.
5. That a Parthenium news letter be produced on a regular basis.
6. That an international workshop be held within three years and continued on regular basis.

Parthenium Networking

In a special session of IPAN, the research workers of Parthenium and weed control unanimously agreed upon and felt the need for the formation of an international network.

The discussion was initiated by highlighting the need for the formation of an international network for effective coordination, planning and implementation of research and extension activities pertaining to Parthenium management. It was felt that countries like India and Australia, where the problem of Parthenium is acute, should take lead in the matter. Participants from different organizations expressed their interest in becoming part of the network. It was unanimously agreed upon to locate the headquarters of International Parthenium Network (IPAN) in India. The recommendations of the conference were finalized in this session and the chairman of the session Dr. Steve Adkins presented the recommendations of the conference which were unanimously approved. He felt the need for creating public awareness through extension agencies and community campaign. He drew attention of the scientists to look for alternative uses of Parthenium and advised not to depend on the sporadic results on its use. He expressed that individually solutions can not be found unless we sit across the table. Thus he felt the necessity of a 'Network' involving different countries. A steering committee should be formed and MOU of various countries be signed to strengthen the research base on Parthenium. He advocated interdisciplinary and inter-institutional approach for the management. He also stressed holding of the second conference in India, the decision about which was already taken in the previous session to hold it in Bangalore, India. He also assured financial support from the ICAR, for the network programme to solve the problem on global basis. This aspect needs to be explored by all concerned.

Second International Conference

The Second International Conference which was held in Bangalore, in India from December 5-7, 2005 made the following recommendations.

Theme 1. Global View of Parthenium

Recommendations

- Severity of Parthenium infestation needs to be reduced to restore ecological balance.
- Chemical weed control methods should be restricted as last resort.
- People should be educated about the ill-effects of Parthenium and its management.
- Research and demonstration in biological control should be intensified.
- Economic losses due to Parthenium infestation and benefits of its control need to be assessed.
- Research efforts on integrated management and utility of Parthenium needs to be promoted.

Theme 2. Importance of Parthenium in Agriculture, Health, Environment and Biodiversity

Recommendations

- Research on the ability of Parthenium to adapt to different tropical eco-systems needs to be intensified.
- Research on plant physiology to study the ability of Parthenium to survive under nutrient-poor environment needs to be initiated.

Theme 3(a). Competitive Plants and Chemical Control

Recommendations

- More emphasis is given to research on management of Parthenium in public lands.

- Use of chemicals to control Parthenium to be discouraged.
- Research focus should be on bio-agents to suppress Parthenium.
- "Integrated Parthenium Management" package to be developed.

Theme 3(b). Biological Control-Insects and Pathogens

Recommendations

- Different institutions, both formal and informal, need to be involved in the release of biological control agents of Parthenium.
- Impact of Parthenium eradication on soil properties need to be studied.
- A detailed data-base on the potential microbial agents needs to be created.
- Host-specific bio-control agents from other parts of the world may be introduced for suppressing Parthenium.

Theme 4. Utility of Parthenium-Medicine, Compost and Insecticidal Properties

Recommendations

- Therapeutic values of Parthenium need to be exploited.
- Parthenium biomass could be silaged and used as animal feed with no detrimental effect.
- Parthenium could be used as a source of compost.
- Insecticide property of Parthenium plant extracts need to be studied.

Theme 5. Course of Action the Formation of International Working Group on Parthenium

Recommendations

- An International Working Group on Parthenium may be formed to coordinate global efforts on Parthenium management.
- Data-base on the present status of Parthenium in the country need to be created.
- National strategy for implementing "Practical Action plan" has to be developed.
- "Global Awareness Campaign" may be launched by declaring year 2006 as "International Parthenium Awareness Year".
- A website containing all the data-base research information on Parthenium may be launched.
- The Australian success model in controlling Parthenium needs to be adopted by other countries.
- Parthenium awareness topics can be included in the school syllabus.
- Efforts need to be made to create awareness among policy-makers and administrators on the importance of controlling Parthenium.

Ten

SCOPE FOR FURTHER RESEARCH

The problem of unwieldy growth of Parthenium in different regions of the globe can not be taken as completely solved with the adoption of the control measures currently being recommended. The integrated Parthenium management technique recently adopted successfully also does not guarantee elimination of this weed, nor is it necessary to eliminate any plant species from the habitat where it survived for a while. The methods which are successful today at one place may or may not necessarily be successful in certain other environments. Therefore, research has to be continued to develop more and more efficient methods applicable to different agro-climatic and soil conditions. Another way of looking at the problem is to see whether the Parthenium plant can be used to meet societal needs. It is reported that it can be ensilaged and the fibrous property of the plant can be exploited for paper and cardboard manufacturing. There are reports to indicate that the plant can be used as green manure. The earlier findings have revealed that this plant has the medicinal properties and chemical constituents that may be useful in manufacturing insecticides and anti-cancer drugs. These aspects are worthy of research.

There appears to be exaggeration in the reports on the spread of this weed across the country. A survey would reveal the reality and it is felt that the recorded Parthenium infested areas is an over estimate. Therefore, a survey might help in getting the correct picture. There is also an exaggeration on the ill-effects of this weed. So there is a need to conduct a survey on the ill-effects on man, livestock, and other vegetation and on the individual ailments reported.

The various parts of *C. sericea* whose plant leachates have shown allelopathic effect on Parthenium is worth investigating to understand the exact nature of the compounds responsible for adverse effect on the germination and growth of Parthenium. Further, the different parts of the *Cassia sericea* plant viz., root, stem, leaf, flower, seeds etc. are likely to find uses for different purposes like tender leaves as leafy vegetables, dry stem as fuel and mulch, seeds for manufacturing gum and extracting proteins which are rich in lysine need to be researched. The very efficient nature of *C. sericea* to harness the solar energy is a plus point since it can produce very high amount of biomass in a given area and time.

There is much controversy about the beetle *Z. bicolorata* which feeds on crops like sunflower, redgram, mulberry, tomato, egg plant (brinjal) and tulsi. Some scientists claim that this insect nibbles only the young leaves or buds, while some others claim that it definitely feeds on these plants, if the Parthenium plants are not available. Again, the point raised is that nibbling itself is quite a serious factor if the insect turns out to be a vector to any disease causing organisms. However these are some of the points that need attention of the researchers interested in validation of true picture.

Parthenium has already invaded residential, unattended lands, cropped areas and, if neglected, would be a menace for decades to come. It is time now for the scientists, administrators and public to take appropriate measures to check it if not eradicate before it assumes monstrous proportions and cause damage to humans, animals and the environment.

Future Strategy and Options

Creating awareness among all the stakeholders in the society is the first step for effective containment of this weed. At present, majority of the people in the society are not aware about Parthenium and its threat to the environment, biodiversity and health of human and animals. Consequent upon this lack of awareness, the control and management of this weed still remains limited to few places, where it is still done on its merit as a regular weed, but not because of its obnoxious nature. In view of the relatively lesser degree of infestation in terms of area, compared to some other states of the country, awareness programme may be clubbed together with suitable containment or preventive or eradication programme specific to different locations and sectors. Following few points are perceived, during this study, for efficient control and management of Parthenium in the region and may be considered in formulation of programmes, policy making and overall management:

- Mapping of Parthenium infested area
- Large-scale demonstration with people's participation in different districts of the state.
- Publication and distribution of leaflets in vernacular language.
- Involvement of mass media for regular broadcasting/publication of information on Parthenium management and their spread.
- Regular monitoring and eradication of Parthenium in newly infested area.
- Proper quarantine measures are to be taken to prevent the infestation of weed in new area.

- Possibility of utilizing Parthenium plants needs to be studied.
- Lectures/demonstration programme on Parthenium control are to be organized among School/College students.
- Control of Parthenium is to be ensured through municipal corporation, gram panchyat, development authority, and railway and other transport organizations.

Eleven

STATUS OF PARTHENIUM GROWTH IN KARNATAKA AND NEIGHBOURING STATES

A recent survey conducted by the author and co-workers on the prevalence of Parthenium in the northern parts of Karnataka *viz.*, Belgaum, Dharwad, Haveri, Gadag, Raichur, Gulbarga, Bijapur, Bidar, Bellary and Koppal districts revealed that compared to its growth in the early 80s as benchmark, Parthenium intensity has been reduced to more than 60% both in agricultural and waste lands. Agents responsible for this control are mostly botanical species with some significant effect here and there of the beetle *Z. bicolorata* in rainy season. Predominant botanical agents responsible in the order of importance are: *C. sericea, C. tora, H. suaveolens* and *Amaranathus spinosus*. In the district of Bijapur in rocky area, however, *T. purpuria* has played a major role. In Uttara Kannada district and also adjoining hilly regions of Belgaum and Shimoga districts, *C. occidentalis* and *Croton sparciflorus* (Plate 31) have played a major role. The two species *viz.*, *C. tora* and *C. occidentalis* have been resisting the entry of this weed in hilly and semi hilly regions even from the initial stages of Parthenium entry into the state. Coming to the southern parts of Karnataka, in the dry regions of the districts of Chitradurga and Chikkamagalur, *Tephrosia purpuria* and *C. occidentalis* have played a major role in preventing the spread of Parthenium. It may not be out of place to mention here that fresh growth of Parthenium is seen in newly opened up forests in the transitional areas and more so towards plains indicating that the weed finds it difficult to establish in the hilly region but it easily establishes in the plains. Also in the plantations of mango, sapota, coconut and such other perennials, the weed puts up luxurious growth if frequent ploughing, intercultivation and weeding are not attended to.

In Bangalore city, vacant sites in many residential areas (Plate 32) where botanical agents like *C. sericea*, *M. jalapa* and *A. spinosus* were planted during early 90s are now free from Parthenium infestation to the extent of 90%. On roadsides, where seeds of botanical agents were sown during early 90s, *C. sericea* replaced Parthenium very fast but in the process of widening of Bangalore – Hosur road, the treated area is all covered under road and nothing can be seen now. However, near Titan watch Company area and Leyland factories near Hosur, replacement of Parthenium by *C. sericea* is quite visible. Frequent disturbance in annual road cleaning activity has deterred the growth of *C. sericea* and other botanical agents. The public and highway authorities need to be educated on the role of naturally existing or deliberately planted botanical agents in controlling the invasion and subsequent growth of Parthenium. On Mysore road in Bangalore city on either side, *Amaranthus spinosus* and *Hyptis suaveolens* have totally taken over by suppressing Parthenium weed. On the old Madras road in the outskirts of Bangalore city *viz.*, on either side of the railway track (near over bridge), replacement of Parthenium by *C. sericea* is distinctly visible. Next to this on the tank bund to the west of high way towards Hosakote, *Hyptis* has replaced Parthenium to the extent of 90%.

In Mysore city, in Kuvempunagar and Nanjangud road, *C. sericea* has replaced Parthenium in short stretches wherever *C. sericea* was sown during 1994–95. In Chamarajanagar district along the state highway between Nanjanagud and Chamarajnagar towns, vast stretches of Parthenium infested road sides are replaced by *C. sericea* particularly near Mahadevapura and Bendaravadi villages. In Chamarajnagar town, *C. sericea* has replaced Parthenium in areas adjacent to Court, First Grade College and Department of Sericulture. Near Srirangapatna along both sides of highway, *T. purpuria* has resisted entry of Parthenium.

Due to efforts of voluntary organizations, sowing of *C. sericea* seeds has started replacing Parthenium very fast in Davanagere city and peripheral areas. In Shimoga district also replacement of Parthenium by *C. sericea* is quite visible on the Harihar road and vacant lands in the city.

In Sangli, Kolhapur and Solhapur districts in Maharashtra state, *C. sericea* species (sericea and to some extent tora) have replaced Parthenium to more than 60%.

In Tamil Nadu, through the efforts of one Captain Kasirajan of Virudhanagar in Madurai district, a stretch of more than one km on the Madurai-Kanyakumari highway and some parts within Virudhanagar town, Parthenium is replaced by *C. sericea*.

In Andhra Pradesh, on the Hyderabad road near Patancheru, some seeds of *C. sericea* sprinkled during 1994–95 have grown and have started replacing Parthenium and the *C. sericea* colonies are enlarging year by year as observed by the author between ICRISAT and Hyderabad city during 2002.

Effect of Beetle

The beetle, *Z. bicolorata* which was released in mid-eighties has spread on its own from single release at

Bangalore and also through separate releases in other parts of the country has spread in many places in Karnataka, Tamil Nadu, Andhra Pradesh and Maharashtra. The beetle has been defoliating the Parthenium plant during the rainy season. Total defoliation leaving only stem is seen in the months of August, September and October in Karnataka and Maharashtra and during October and November in Tamil Nadu. Suppression of Parthenium during the rainy season is very impressive and the release of *Z. bicolorata* can serve as an important component of the integrated Parthenium weed management technology. Australian efforts in educating public on biocontrol approach are given below.

Tips for establishing biological control– Australian model

- Establish suitable nursery sites. An ideal site should have constant moisture, lush growing *Parthenium* and a continual replenishment of biological agents
- Biological control agents are not 'sprays'. So, consider their location of action carefully
- Release at an appropriate stage in *Parthenium* life cycle
- Successful control often results from using a combination of various biological control agents
- Biological control agents may fail to establish in some areas, even-after well-planned collection and distribution. So, try, try and try again!

Conclusion

The onus now on us is as to how to assess the human, educational problems and bring forth the knowledge to the people and agencies concerned to oversee the impending measures for Parthenium control.

It may be said that still there is a huge hidden information on Parthenium existence/perpetuation in India and elsewhere. It is possible that a sustained and satisfactory progress from all sectors will entail education of people and with research appropriate attention and funds are made available. Then only, hopefully, we shall be able to tackle the problem, with more ease and success in a short period.

The formation of action groups on Parthenium such as Philosophy & Social Action, Society for Parthenium Management (SOPAM), Parthenium Action Group Inc. and others is essential at local, state, country and international levels. We must assess the impact of these programs. The multidisciplinary research and appropriate transfer of technology from laboratory to field applications is extremely important for managing this invasive plant. The significance of Parthenium in today's environment is well established. There is no doubt about the potential detrimental impact of this plant on human affairs and the environment. We must continue our research and public education on this invasive species.

Twelve

FREQUENTLY ASKED QUESTIONS

The Parthenium has attracted the attention of people from all walks of life for it has been known to adversely affect all living beings and pollute the environment with fast invading and perpetuating character, defying the law of natural plant succession. There have been ever pouring questions about the entry, growth habit, harmful effects and utility of this plant. Therefore, it is thought necessary to compile all the questions that have been posed to research workers and environmentalists on this plant put under a separate head.

1. What is so called Parthenium and where is it seen?

It is an invasive annual pernicious plant seen in many countries introduced accidentally along with imported grains. Although it is seen in many countries, its growth intensity is not in dangerous level in its native countries. It is very rampant in newer environments like in India, Australia, Ethiopia, Nigeria, Bangla Desh, Pakistan, Sri Lanka, etc.

2. How Parthenium affects agriculture and environment?

Because of its invasive capacity with very strong allelopathic effect and adaptability to a variety of agro-ecological situations and hardy nature, it is capable of invading all kinds of ecological situations and suppresses or even causes extinction of native flora. Because of its allergic pollen, leaves and hairs it causes skin problems and aggravates many skin as well as nasal problems in sensitive individuals. It invades all kinds of unattended lands and reduces food and fodder production. Its pollen load in atmosphere is dangerous to human beings as well as animals. It suppresses the normal growth of plantation crops, since such crops need 3 to 6 years to establish with full canopy.

3. What is the extent of damage caused by Parthenium to human beings, animals and agriculture?

No definite estimates are made in respect of the damage caused to human beings, animals and agriculture. But there are studies indicating that, in the peak period of its growth in rainy season, about 40% of the skin patients going to hospitals were due to allergic reaction to this weed. Animals are affected in different ways like lesions in different parts of the body, milk getting poisoned with "parthenin" when the cattle grazes the grass etc around the Parthenium colonies. Its invasion reduces grazing area as well as the production of fodder. In fact, it is not a serious problem in agricultural lands where clean cultivation is practiced. But in young orchards and fallow lands, it is a serious problem. Under neglected cultivation, the yield losses due to this weed are reported up to 70 %.

4. Where in India it is seen and its extent on different locations?

It is seen in almost all states of India except in interior forests and coastal districts of Karnataka, Kerala, Goa and Maharashtra.

5. Is it useful to human beings and domestic animals in any way?

It is reported to have anti-cancer properties and also insecticidal properties but it is yet to be exploited. It is now being used for compost making. It is being mixed with flowers in preparing bouquets, garlands, decorative items etc. not having knowledge of its harmful effects. It is also being used as centering material to form base for casting cement concretes by small contractors.

6. Which parts of the plant are harmful?

Pollen is most harmful. Even hairs on stem and leaves, and leaves themselves also cause allergic reactions. Allelopathic effect causing suppression of natural flora is due to leachates of the entire plant getting into soil after the plant completes its life cycle and dries up. It is reported that seed and pod leachates have stronger allelopathic effect.

7. How does this weed spread?

It spreads through all seed dispersal agents such as air, water, animals, human beings, any moving objects such as automobiles, insects etc.

8. What are the methods of controlling this weed?

There are manual, chemical, cultural, utilization, biological and integrated methods used but the most effective and eco friendly method is the integrated method which involves all these methods as components depending on the type of land and season etc, But, major components are biocontrol agents supplemented by mechanical removal before flowering.

9. What is integrated Parthenium weed management (IPWM)?

The major component of this method is use of biocontrol agents such as plants, parasites such as insects and pathogens supplemented with cultural practices, manual removal and herbicides. In India, it is botanical agents so far; but in Australia it is insect agents. But chemicals are recommended as a component only as a last resort where emergencies arise for getting rid of this weed for specific reasons.

10. What are bio-agents and how this weed can be controlled by bio-agents?

Bioagents can be plant species, insects; microbes, virus etc. It is allelopathic mechanism in botanical agents, and feeding on different parts of the plant by insect agents and infestation in case of pathogens. The botanical agents are many like *Cassia* spp., *Croton sparciflorus*. *Hyptis* spp., *Sida* spp., *Tephrosea* spp., *Amarnathus* spp. etc which are described in detail in this book. Insect agents employed mainly in Australia and *Zygogramma bicolorata*, a defoliator used in India can be cited here. Use of microbes is yet to be developed for field adoption.

11. What is the present status of research on Parthenium management?

It is not systematically taken up anywhere. It is going on in a fragmented manner in different countries. There is a need for coordination and collaboration to achieve quick results. Two international conferences (Dharwad, 1997; Bangalore, 2005) have taken place, and the third one is going to be held in Mysore during September, 2009. The recommendations are to be implemented both for suppressing the spread of this weed and also by exploiting its utility/economic value.

12. Are there situations where this weed does not grow?

Yes. In coastal laterite soils with heavy rainfall, thick forests and places where natural enemies prevail. It is not seen in regions where natural biodiversity is not disturbed. In India, it is not a problem in the west coast and in thick unopened forests.

13. It is stated that Parthenium has defied natural law of plant succession? What does it mean?

No single weed will be seen eternally in any waste or unattended land. There will be one or more weeds replaced periodically by one or more weeds in the natural course. Normally a new weed thrives for 10 to 20 years after its invasion and gives way to the successive members. This is termed as natural succession. But Parthenium is recalcitrant, and, in India, it has not allowed any successor for the last 5 to 6 decades. Thus, this weed has definitely defied natural law of succession.

14. How many beetles should be released?

One adult was found to bring defoliation of a single Parthenium plant in 6–8 weeks. Theoretically, if releases are to be carried out at this rate, about 0.4–0.7 million insects will be required per hectare, as the weed density varies from 40 to 70 plants per square meter. In practice, it is neither possible nor necessary to release so many insects as they are capable of multiplying rapidly. The reports of National Research Centre for Weed Science recommends that releases of about 500–1000 beetles can bring about establishment and eventually show the expected results of defoliating the affected Parthenium population. Once plants are eaten up in the release spot the insects migrate into adjacent areas. Taking this into consideration, a number of release spots can be selected in a particular place or city, which can act as a focal point. More releases mean quicker establishment of the beetle. Therefore, for better control, do as many releases as affordable during first couple of years of introduction. This method reduces the time in years for the beetle to build up population and helps the beetles to disperse fast. The least affordable approach is to introduce one or two releases into infested area and do nothing more. This method will get a colony started, but will be slow in terms of time and area.

15. What happens to the insects after the weeds are eaten?

Parthenium will never be eradicated in a vast country like India. Some plants will always escape from the attack, which will allow the insect population to sustain itself during years of low weed density.

16. What aspects of management of Parthenium could be easily initiated by common people? How to propel the programme?

The first thing is to educate the people about its harmful effects to draw the attention of people concerned. With some education on the integrated method, common people can get directly into action by collecting seeds of botanical agents, break seed dormancy and wait for the next rainy season to take up sowing. Also try to educate public that there are natural enemies among wasteland weeds which, if protected and their growth promoted, will not allow Parthenium to invade. The fresh flushes of Parthenium plants that keep coming up at frequent intervals need to be uprooted before flowering and composted. More research is also needed to simplify the technology towards easy adoption.

17. Can the management practices of Parthenium be entrepreneurial or income generative?

Yes. After gaining enough experience in practicing IPWM, contractual services can be extended to needy institutions, municipalities and corporations to take up Parthenium control programme on agreed cost for agreed period. This can fetch profits to groups and can be a livelihood.

18. What is the existing mechanism for popularizing the ill-effects of Parthenium and its control, what improvements need to be made?

At present there is no systematic way of segregating the patients affected by Parthenium allergy and treating them. Special cells, to work during the peak period of Parthenium growth, to treat the patients and create awareness in the public in order to avoid coming in contact with Parthenium.

19. What role institutions, universities and NGOs play to convey the message on Parthenium ill-effects to educate and enlighten community, literate and semi-literate persons and illiterate farmers?

It is important to educate first on the ill-effects of this plant through various media, besides designing awareness campaigns and holding them at appropriate time. Once they are convinced about the harmful effects, they will start listening as to how to suppress the growth of this weed. There is integrated method successfully adopted for certain situations where research is carried in the past 2 to 3 decades. But there is no well defined single universally applicable management technology to contain the growth of Parthenium in all eco-systems. But, the integrated method developed and adopted here and there could be fine tuned with adaptive research to suit different agro-climatic situations before the technology is successfully adopted. Institutions and NGOs can take up this task. Also NGOs and enlightened communities can see the areas where Parthenium has been successfully controlled and educate the people around them and create confidence that this weed can be controlled. The advantage of community approach to fight against this weed is very important. In view of the fact that "It has high seed fecundity" and that "it cannot be achieved by individual approach" needs to be brought home.

20. Can Parthenium management be made as a rural enterprise?

There is unemployment problem in rural areas for school and college drop outs. Even many graduates are unemployed. They are capable of understanding the technology, and science behind the technology, and hence more capable of convincing others and harnessing the cooperation of all. They can form groups, enter into contract with institutions and government departments and private sectors to reclaim their premises and activity areas to free them from Parthenium invasion and can have time based agreement. This can become a livelihood also. It is easy to employ laborers and collect seeds of botanical agents, dry them, break dormancy and use them after first showers. Seeds can also be sold to needy organizations.

21. How to use the student force of schools in Parthenium control-both in urban, semi urban and rural areas?

At present, NSS groups are being used to adopt different methods, more of hand pulling and destruction. It is universally known in India that there are sensitive people and they are to be excluded in such operations and all those who are engaged in pulling operation are provided with nose gags.

22. Can Parthenium control education be linked with adult literacy programmes?

Yes. Any educational activity can get this subject added into their curriculum and create interest among the people involved in such activities. It is very fitting subject to be included under education on environment.

23. Are there ayurvedic and homeo drugs for Parthenium allergy control besides allopathic drugs?

I am not aware of these aspects. I have been concentrating mostly on management aspect.

24. What is the practical vision on Parthenium management in the next decade or so?

The environmentalists, policy makers and public are to be educated on the ill-effects and need for developing situation specific management technology along with adequate support. It is hoped that day will come.

25. What is the economics of Parthenium control at present?

It is very subjective and difficult to answer this question. The intensity of invasion and growth, season, seed productivity of the botanical agents chosen, labour wages prevailing in the area intended for treatment for seed collection and preparation count in this aspect.

Thirteen

ASSORTED TOPICS ON PARTHENIUM

Even though the book is designed to bring out all information generated by well known scientists on Parthenium from all over the world, there is still an abandant literature which could not be fitted into the various chapters of the book. Such valuable information mainly toxicology, chemical aspects, etc. is provided in this chapter for the benefit of the readers.

Invasive Species

An invasive species is a non-native species whose introduction does, or is likely to, cause economic or environmental harm or harm to human health. An invasive species can be a plant, animal, or any other biologically viable species that enters an ecosystem beyond its native range.

Environmental Weeds

Environmental weeds are introduced, aggressive species that colonize natural vegetation and suppress the native species to a certain extent. In Europe, *Robinia pseudacacia, Prunus serotina, Amelanchier* species are considered as environmental weed trees. *Elodea canadensis* and *Elodea nuttallii* are two aquatic environmental weeds. In Nepal, *Parthenium hysterophorus* can be taken as an example of this type as it dislodges mainly *Cassia tora* in the *Terai* and *Inner-terai* and *Cannabis sativa* and *Artemisia species* in the Kathmandu valley.

Allelopathy

The term means the injurious effect of one upon another, it was coined for the production of chemical compounds (allelochemicals) by one plant, mostly secondary metabolites, that can induce suffering or give benefit to, another plant. The role of allelopathy may be stimulatory or inhibitory. In an agro-ecosystem, allelopathic interactions may be crop-crop, weed-weed and crop-weed. If the allelochemicals of one weed inhibit germination, growth, and development of another weed, they can be called as natural herbicides. A crop may be stimulatory to growth and development of other crop. Such crops with stimulatory potential can be used as intercrops or as rotational crops. A large amount of data is accumulating on allelopathy.

Beneficial Effects for Agriculture

We have to look weeds from an ecological point of view. For instance, in an agro-ecosystem, weeds not only compete with crops but also compete with their own population. If we control weeds by herbicides, it results in resistant weeds which only compete with crops. Because of this reason Austrian Professor Holzner proposed his idea about ecological management saying "it would seem reasonable and subject to aim at a weed control achieving a weed population in the fields that is rich in species but poor in individuals and that can be easily controlled, mainly by mechanical means and by crop rotation, and does not tend to the negative effects of compensation. "In other words" not total eradication of weeds but management of the weed population, using tolerable species to control other more noxious ones....".

Parthenium Taxonomic Affinities

Within the genus *Parthenium*, only one species, *Parthenium argentatum* Gray (guayule), is of potential economic value. Guayule has frequently been suggested as potential source of natural rubber, but has never been commercially exploited. *P. argentatum* is chemically and morphologically very different to *P. hysterophorus* (Stuessy, 1977) and their insect and disease complexes are also different (McClay *et al.*, 1995).

Parthenium Pest Status

In Queensland, Parthenium weed is a P2 declared plant throughout the state except in designated areas, where it is a P3 or P4 category plant (Anon., 1997). It is a declared noxious plant in New South Wales and Victoria, a Class A and C plant in the Northern territory, and a P1 plant in Western Australia.

Parthenium is an invasive annual of open land and pastures. It is unpalatable to stock, and on suitable soils in the summer rainfall tropics and sub-tropics, will quickly dominate both native and planted pastures particularly where these are overgrazed. Once dominant, Parthenium continues to persist as pure stands unless managed. Plants are capable of flowering when one month old and will remain in flower for 6–8 months when conditions are suitable.

In central Queensland, Parthenium weed quickly invades disturbed soils such as overgrazed pastures and newly cleared or eroded lands, where it remains dominant unless the appropriate management techniques are applied. Management requires de-stocking to allow grass growth, followed by the reduction of stocking rates by about 40% to prevent re-invasion.

Because one of the major methods of spread into new areas is via infested pasture seed, glaziers have become increasingly unwilling to purchase seed from areas infested with Parthenium weed, and in 1989 the Queensland Department of Primary Industries introduced a new inspection and sampling scheme by which growers can obtain a certificate that seed is free of Parthenium weed. The scheme costs the grower about $70 per ton but has been widely used by seed growers from central Queensland. Special facilities to clean headers and harvesting machinery have been set up at border areas to reduce spread into NSW. Sale prices of cattle properties in central Queensland are being discounted by up to 30% because of Parthenium weed infestations. Sale of grain from central Queensland to countries in South-East Asia have been affected by fears of contamination.

Ingestion of the plant produces unacceptable (Tudor *et at.*, 1982). The plant causes acute allergic eczematous dermatitis in humans, which under continued exposure become chronic. Highly sensitised individuals are forced to avoid contact with the plant entirely. In some situations, landholders and other workers are forced to move from their normal activities.

Chippendale, J.F. and Panetta, F.D. 1994. The cost of Parthenium weeds to the Queensland cattle industry. *Plant Protection Quarterly* 9: 73–76.

In Queensland, Parthenium weed commonly dominates cultivated and other disturbed areas, in addition to flood-prone pastures. The presence of Parthenium in cropped lands can almost double cultivation costs and restricts the sale and movement of contaminated produce. In 1990–

91, a mail survey of beef producers was conducted in the most heavily infested region in central Queensland. Annual losses caused by this weed were found to be in the vicinity of $ 16.5 m. Losses comprised opportunity costs (*e.g.* reduced stock numbers and liveweight gains), as well as additional production and control costs. Increased expenditure on research into Parthenium control (especially biological control, for which research expenditure was approximately $ 3,50,000 during 1990–1991) is thus warranted.

Navie, S.C., McFadyen, R.E., Panetta, F.D. and Adkins, S.W. 1996. A Comparison of the Growth and Phenology of two Introduced Biotypes of *Parthenium hysterophorus. In*: Proc. 11th Aust. Weeds Conf., R.C.H. Shepherd (*ed.*) Weed Sci. Soc. Victoria, Frankston. pp. 313–316.

Two biotypes of *Parthenium hysterophorus* L. are established in Australia as a result of two separate introductions from the USA. The first introduction occurred in south-east Queensland and the second in central Queensland. Nine plants from each of the biotypes were grown under a day/night temperature regime of 23/13°C and 14.5 h photoperiod in a plant growth cabinet for a period of five months. Plants from the central Queensland biotype had a higher dry weight production, earlier stem elongation, were taller; produced larger diaspores and had a lower percentage of filled seed than the plants from the south-east Queensland population.

These differences in biology may offer an explanation why the central Queensland biotype is very aggressive while the south-east Queensland biotype is relatively well contained.

Navie, S.C., McFadyen, RE., Panetta, F.D. and Adkins, S.W. 1996. The biology of Australian weeds. *Parthenium hysterophorus* L. *Plant Proto Quart.* 11: 76–88.

This review contains a general overview of Parthenium and the problems it causes, and includes details on distribution, reproduction, hybrids and population dynamics. A discussion on the management of *Parthenium hysterophorus* is also provided.

Biological Control Abstracts

Dhileepan, K., Madigan, B., Vitelli, M., McFadyen, R.E., Webster, K. and Trevino, M. 1996. A new initiative in the biological control of Parthenium. *In*: Proc. 11th Aust. Weeds Com., R.C.H. Shepherd (*ed.*) Weed Sci. Soc. Victoria, Frankston. pp. 309–312.

Parthenium is an extremely prolific weed and causes severe economic loss, health problems and habitat destruction in Queensland. Biological control of the weed is the only cost-effective, environmentally safe and ecologically viable method available. Eight species of exotic insects have been introduced into Australia for the biological control of Parthenium weed, of which at least six species are known to be established. The stem galling moth *Epiblema strenuana,* the stem-boring weevil *Listronotus setosipennis,* and the leaf-mining moth *Bucculatrix parthenica* are the species that are successfully established in most of the Parthenium infested areas of Queensland, while the leaf feeding beetle *Zygograrama bicolorata* and the seed-feeding weevil, *Smicronyx lutulentus* appear to be established only in the central Queensland region.

The new initiative program on the biological control of Parthenium is currently mapping the distribution of various biocontrol agents in Queensland, and assessing the role of already established biocontrol agents in controlling the weed populations. The stem boring weevil *Conotrachelus* sp. and the stem-boring moth *Platphalonidia mystica* both from Argentina are being released in the central Queensland region. In addition, two new root-feeders, *Carmenta ithecae* and *Thecesternus hirsutus,* which are being imported from Mexico will be host-tested. The newly formed Parthenium Action Group is involved in the distribution and field evaluation of already established biocontrol agents in the central and northern Queensland regions.

McClay, A.S., Palmer, W.A., Bennett, F.D. and Pullen, K.R. 1995. Phytophagous arthropods associated with *Parthenium hysterophorus* (Asteraceae) in North America. *Environ. Entomol.* 24: 796–809.

A faunal survey was conducted to find possible biological control agents for *Parthenium hysterophorus* L. (Asteraceae), a serious annual weed in many parts of the world, particularly Queensland, Australia, and India. Between 1977 and 1991, most of the plant's native range in North America was surveyed from bases at Monterrey and Cuernavaca, Mexico, and Temple, TX. Two hundred and sixty-two phytophagous orthropod species were collected on *P. hysterophorus* by various methods including hand picking, dissection, rearing, and sweeping; 144 of these species were found to feed on the plant at some stage of their life cycle. The orders represented most abundantly were: Coleoptera (33.2%), Homoptera (22.9%), Lepidoptera (20.2%), and Hemiptera (18.3%).

Two fungal pathogens, Dietel and Holway variety *partheniicola* (Jackson) Parmelee and *Puccinia melampodii* Dictel and Holway, were also observed. An index of similarity was used to make pairwise comparisons between the phytophagous arthropod communities on different plant taxa. These comparisons showed that the fauna of *P. hysterophorus* is most similar to that of ragweeds (*Ambrosia* spp.). Six insect species that were shown to be stenophagous were shipped to Australia for further testing and possible field release. One of the fungal pathogens underwent host-range testing in the United Kingdom and was released in Australia.

McFadyen, R.E. and Cruttwell. 1992. Biological control against Parthenium weed in Australia. *Crop Protection* 11: 400–407.

Parthenium hysterophorus is an annual herbaceous plant native to the tropical Americas, which is widely adventitious and does not occur in East Africa, parts of Asia and Australia. It is a major crop and pasture weed in India and Australia in particular, where it is also a human health hazard, causing allergic dermatitis and respiratory problems. A biological control programme involving the introduction of insects from the Americas started in Australia in 1975 and is still in progress. Six species of insects have been released, of which four are established but only one, the moth *Epiblema strenuana*, is exerting significant control on the weed. The moth larvae form galls in the Parthenium stems and shoots; damage by several larvae stunt plants and reduces seed production.

Unfortunately, the erratic climate interferes with the control; long dry periods reduce the moth population to very low levels so that when the Parthenium germinates after rain there is inadequate control. Efforts to establish other biocontrol agents are continuing in Australia. India rejected *E. strenuana* because of attack on *Guizotia abyssinica* in tests, but has successfully established the leaf-feeding chrysomelid *Zygogramma bicolorata* with promising results. Other insects evaluated in Australia are available and should be tested for use in India and other countries with the Parthenium problem.

Health Abstracts

McFadyen, R.E. 1995. Parthenium weed and human health in Queensland. *Australian Family Physician* 24: 1455–58.

Allergic reactions to the pollen and plant dust of the Parthenium weed are causing major health problems in central Queensland, which are expected to increase, especially as the pasture weed is rapidly spreading south. This paper reviews published information on health aspects of this weed and calls attention to its spread into areas with much greater population.

Weeds – A source of medicinal plants in ayurdeda and it's therapeutic applications, with a special note on ayurvedic treatment for Parthenium allergy

Ayurveda considers that there is nothing on the earth which is not a medicine and this substantiates, weed plants as source of medicines. *Parthenium hysterophorus,* basically not of an Indian origin, but introduced to India in 1955 causes severe human and animal health problems, known for its skin and respiratory allergy manifestations. Here is an attempt to describe the ayurvedic treatment of Parthenium allergy by correlating to sheetapitta, udarda, dooshivisha, oshadhigandhaja jwara and adopting basic principles of ayurvedic therapeutics namely samshodhana (elimination therapy), samshamana palliative-internal and external medicaments) which is yielding better results, has to be evaluated by scientific study. This may be used as green leaf manure, as a biopesticide and as compost to crop plants apart from utilizing it as a drug resource from recent studies. In *Ayurveda* classics nearly more than sixty-nine weeds, have been described as medicinal plants with its therapeutic applications. With advancement of science, which enables the quick isolation, purification and characterization of compounds and their evaluation against large specific screen, based on cell based mechanisms. While this approach holds great promise for the discovery of NCEs (new chemical entities), that may become useful drugs. It totally ignores the *ayurvedic* basis for prescribing the herbal preparations in the first place.

Management of Parthenium Allery in Ayurveda

By adopting the above principles, the following treatment can be adopted: the swedana (fomentation), vamana (emesis) and virechana (purgation) followed with administration of following ayurvedic medication internally (Sushruta samhitha). They are:

1. Dooshi visha agada: 48 g as decoction to be given frequently (Sushruta samhitha).
2. Amrita ghrita: 12 g twice daily in empty stomach with hot milk or hot water (Priyavrat Sharma)
3. Haridra khanda: 6 g thrice daily after food with either milk or water
4. Gulgulu tikthaka kwata: 48 g as decoction twice daily in empty stomach
5. Sootha shekara rasa: 125 to 250 mg thrice daily after food with ghee or honey

Gopalan, C., Sastri, B.V.R. and Balasubramanyam, S.C. 1971. *Nutritive value of Indian foods*. National Institute of Nutrition, FCMR, Hyderabad. pp. 36.

Parthenium - an exotic weed as a source of plant nutrient in potato crop

A field experiment was conducted at Main Research Station, GKVK, University of Agricultural Sciences,

Bangalore during *rabi* seasons of 2002 and 2003 to study the effect of alien weeds like *Parthenium hysterophurus, Chromalaena odorata and Lantana camara* as nutrient source on yield and yield attributes of potato. There were 15 treatments replicated thrice. The experiment involved the treatments with substitution of 25 and 50% N through compost and weed biomass incorporation and remaining with inorganic N. These treatments were compared with 100% N as FYM, 100% NPK alone and 100% NPK+10 t FYM/ha. The results revealed that maximum tuber yield of 21.05 t/ha was recorded in 100% NPK integrated with 10t FYM which was statistically on par with substitution of 25% N as *Parthenium hysterophorus* compost (18.27 t/ ha) or *Lantana camara* incorporation (18.20 t/ha). The yield attributes are also in the same order. The effect of different weed utilization treatment was found to be non significant on total weed density and dry weight of weeds. This experiment envisages the possibility of saving 25% of inorganic fertilizer by utilizing Parthenium as compost and chromolaena or lantana as incorporation to potato crop under eastern dry zone of Karnataka, India.

Temporal variation in relative dominance of *Parthenium hysterophorus* and its effect on native bio-diversity

There were twenty-two weed species noticed in the site dominated by dicots followed by monocots. *Parthenium hysterophorus* was the sole dominating species among dicots followed by *Ageratum conozoydes. Cyperus rotundus* and *Cynodon dactylon* were the dominating monocots. The study revealed that there is temporal variation in the relative dominance of Parthenium. There was sharp increase in the relative dominance of Asthenia from June to October where as the absolute density of Parthenium increased from June to August and then started declining. The importance value as well as the summed dominance ratio of Parthenium also increased from June to October.

The native bio-diversity of weeds decreased as the density of Parthenium increased over time. There was considerable reduction in the species richness as the density of Parthenium increased. There was also a sharp decline in the diversity index, evenness and species richness as the density of Parthenium increased over time clearly indicating the threat of Parthenium on the native bio-diversity of weeds.

Extension strategies for integrated management of Parthenium

1. Use well decomposed organic matter to field crops, thus help in reducing the germination of weeds.
2. Restrict the movement of Parthenium from one place to other after weeding and burn them on the spot as for as possible or burry in the soil.
3. Remove the weeds before flowering. This reduces the reproduction. This helps to keep our environment clean and hygienic.
4. Remove the weed plant along the entire root.
5. Remove the weed during rainy season by wearing plastic gloves or covers and also by covering the exposed portion of the body.
6. Use of competitive plants like *Cassia sericea, Cassia tora, Stylozanthus scabra. Croton sparsiflorus* and *Tagetes erecta* for suppression of Parthenium and there is a need to supply seeds of these plants to the farmers at low cost or even free.
7. Release of 500 Mexican beetle *Zygogramma bicolorata* per acre in the Parthenium infested area and to facilitate spread of the beetle to other areas.
8. Use of post emergent herbicides glyphosate @ 8–10 ml/litre of water.

Fourteen

LITERATURE CITED

Adkins, S.W., Navie, S.C. and McFadyen, R.E. 1996. Control of Parthenium weed (*Parthenium hysterphorus* L.): A centre for Tropical Pest Management Team Effort. 11th Australian Weeds Conference, R.C.H. Sheperd (*ed.*) Weed Science Society, Victoria, Frankston, pp. 573–578.

Agashe, S.N. and Prathibha Vinay, 1975 Aeroplynological studies in Bangalore city. I. Pollen morphology of *Parthenium hysterophorus* Linn. *Current Science,* **44**: 216–217.

Aneja, K.R., Dhawan, S.R. and Sharma, A.B. 1991. Deadly weed *Parthenium hysterophorus* L. and its distribution. *Indian J. Weed Sci.,* **23**(3&4): 14–18.

Angiras, N.N. and Saini. J.S. 1997. Distribution, menace and management of *Parthenium hysterophorus* L. in Himachal Pradesh. *In: Proceedings of the first International Conference on Parthenium Management* held at the University of Agricultural Sciences, Dharwad, Karnataka, India. pp. 13–15.

Arny, H.V. 1890. *Journal of Pharmacology*, **62**: 121.

Arny, H.V. 1897. *Parthenium hysterophorus. Journal of Pharmacology,* **69**: 168.

Anonymous, 2005. Biological control of Parthenium through *Zygogramma biocolorata*, published by National Research Centre for Weed Science.

Bala, S.K., Bhattacharaya, P., Mukerjee, K.S. and Sukul, N.C. 1986. Nematicidal properties of plants *Xanthium strumarium* and *P. hysterophorus*. *Environment and Ecology* **4**(1): 139–141.

Banwari Lal, Tomer, P.S. and Prasad, J.V.N.S., 1997. Effect of Parthenium extracts on the germination and seedling growth of oats. *In*: *Proceedings of the First International Conference on Parthenium Management* held at the University of Agricultural sciences, Dharwad, Karnataka, India. pp. 164–165.

Barton, J. 2004. How good are we at predicting the field host-range of fungal pathogens used for classical biological control of weeds? *Biol. Control*, **31**: 99–122.

Bennett, S.S.R., Nathani, H.P. and Raizada, M.B. 1978. *Parthenium hysterophorus* L. in India – A review and history. *Indian J. Forestry* **1**: 128–131.

Bhuti, S.G. and Hiremath, I. G., 1997. Parthenium in the control of shoofly. *In*: *Proceedings of the First International Conference on Parthenium Management* held at the University of Agricultural Sciences, Dharwad, Karnataka, India. pp. 130–132.

Bisht, G. 2004. Experts turn to beetles to check the growth of congress grass. *Indian Express*, June 4, 2004.

Bidhas Ray, 1975. Wage weed war. *Hindustan Times*, Delhi, September 8, 1975.

Butler, J.E. 1984. Longevity of *Parthenium hysterophorus* L. seed in the soil. *Australian Weeds* **3**: 6.

Castex, M.R., Molfino, J.F. and Ruiz Moreno, G. 1940. Primera contributional estudio de la flora alergogena de la Republica Argentina. *Parthenium hysterophorus*; nota previa (First contribution to the study of the allergic flora of the Argentinian Republic, *Parthenium hysterophorus*). *Bol. Acad. Nac. de Med. de. Buenos Aires.*, pp. 91–96.

Chakravarthy, A.K. and Bhat, N.S. 1994. The beetle (*Zygogramma conjuncta* Rogers), an agent for the biological control of weed, *Parthenium hysterophorus* L. in India feeds on sunflower (*Helianthus annuus* L.). *Journal of Oilseed Research*, **11:** 122–125.

Chaktravarthy, A.K., Rajagopal, D. and Bhat, N.S. 1997 Evaluation of suppressive methods and future strategies for the management of *Parthenium hysterophorus* Linn. *In*: *Proceedings of the First International Conference on Parthenium Management* held at the University of Agricultural Sciences, Dharwad, Karnataka, India. pp. 173–176.

Chandras, S. 1970 Paper presented at 'Symposium on problems caused by *Parthenium hysterophorus* in Maharashtra region, India. Marathi Vignayan Parishad, Poona, July 27, 1968 (*eds.*) G.S. Chandras and V.C. Varthak. *PANS* **16**: 212–214.

Chandra, S. 1973. Parthenium – A new dermatitis weed in Bihar. *Proc. Sixteenth Indian Sci. Congress* II. pp. 375.

Char, M.B.S. and Bhat, S.S. 1975a Parthenin – A growth inhibitor behaviour in different organisms. *Separatum Experientia* **31**: 1164.

Char, M.B.S. and Bhat, S.S., 1975b, Antifungal activity of pollen. *Dei Naturwissnschaften,* **62**: S. 536.

Chetti, M.B., Patil, Y.H., Hiremath, S. M. and Srikant Kulkarni. 1997. Allelopathic effect of Parthenium leaf and root extracts on biophysical and biochemical parameters of eupatorium, (*Chromolaena odorata* K & R*)*. *In*: *Proceedings of the First International Conference on Parthenium Management* held at the University of Agricultural Sciences, Dharwad, Karnataka, India pp. 142–148.

Chupp, C. 1956. A monograph of the fungus genus Cercospora. Ltaca, New York. pp. 667.

Ciferri, A. 1956. Microflore domingensis exsiccate. *Sydowia* **10:** 130–180.

Curtis. 1921. *Botanical Magazine,* 49 T. 2275, Edward Couchman, London.

Dale, I.J. 1980. The habitats of *Parthenium hysterophorus* L. in South America, Mexico and the United States of America compared with the Qeensland environment. Alan Fletcher Research Station, Sherwood, Queensland. pp. 14.

Dam, D.P., Dutta, R.M. and Dam, N. 1993. An early collection record of *Parthenium hysterophorus* L. from botanic garden, Calcutta. *J. Bombay Natural His. Soc.* **90**: 128.

Darlington, C.D. and Janaki Ammal, E.K. 1945. *Chromosome atlas of cultivated plants.* George Allen & Unwin Ltd., London.

Das, B., Bhattacharya, A., Sahu, P, K., Chetti, M.B. and Hiremath, S. M. 1997. Water uptake patterns in *Cassia tora* L. as influenced by different seed types. *In: Proceedings of the First International Conference on Parthenium Management* held at the University of Agricultural Sciences, Dharwad, Karnataka, India. pp. 70–73.

Dua, K.L. and Shivpuri, D.N. 1969 Atmospheric pollen studies in Delhi area in 1958–59. *Journal of Allergy*, **33**: 507–512.

Ekanayake, H.M.R.K. 2004. Biological control of *Parthenium hysterophorus.* Report submitted to CARP: 12/471/352, 2002–2004. CARP.

Ellis, J.L. and Swaminathan, M.S. 1969. Notes on some interesting plants from south India–I. *Journal of Bombay Natural History Society,* **66**: 233–234.

Evans, H.C. 1997. The potential of neotropical fungal pathogens as classical biological control agents for management of *Parthenium hysterophorus. Proc. First international Conference on Parthenium Management*, Vol. I.

Evans, H.C., Greaves, M.P. and Watson, A.K. 2001. Fungal biocontrol agents of weeds. *In*: Fungi as Biocontrol Agents (Butti, T. M., C.W. Jackson and N. Magan, *eds.*). CABI Publishing, Wallingford, Oxon., UK. pp. 169–192.

Evans, H.C. 2002. Biological control of weeds, *In*: The Mycota XI: *Agricultural Allications* (Kempken, F. *ed.*) Springer-verlag, Berlin, Germany. pp. 135–152.

FAO. 1996. *International Standards for Pystosanitary Measures.* Secretariat of the International Plant Protection Convention, Rome, Italy.

Farkya, S., Pandey, A.K. and Rajak, R.C. 1996. Mycoherbicide potential of *Fusarium* spp. agent Parthenium: Factors affecting *in vitro* growth and sporulation. *Bioved* **7**: 1–9.

Fauzi, M.T., Tomely, A.J. Dart, P.J., Ogle, H.J. and Adkins, S.W. The rust *Puccinia abrupta* var. *patheniicola*, potential bio-control agent of parthenium weed: Environmental requirements for disease progress. *Biological Control* **14**: 141–145.

Fernandez, E.F. 1942. La cicutilla: fitoterapia autoctona. Exc. Med. Secretaria de Communicacionesy obras publicas, Mexico, 1 (dec. 7), pp. 245–247.

Fernold, M.L. 1970, Gray's *Manual of Botany*. 8th edn. Van Nostrand Reinhold Co. New York.

Fisher, A.A. 1996. Esoteric contact dermatitis. Part IV: Devastating contact dermatitis in India produced by American *Parthenium* weed (the scourge of India). *Cutis New York*, **57**: 297–298.

Gajendra, G. and Gopalan, M. 1982. Notes on the antifeedant activity of *P. hysterophorus* Linn. on *Spodoptera litura* Fabricious (Lipidoptera: Noctuidae). *Indian J. Agric. Sci.* **52**: 203–205.

Gidwani, L. 1975. Weed out congress grass or else face disaster. *The Current*, 25 October, 1957, pp. 12–13 &17.

Gleason, H.A. and Cronquist, A. 1963. *In*: Manual of vascular plants of North-Eastern United States and adjacent Canada, New York. pp. 691.

Govindappa, M.R., Colvin, J., Keshava Murthy, K.V. and Muniyappa, V. 2004. Factors driving tomato leaf curl geminivirus disease epidemics and its implications in south India and management implications. *Proc. Second European Whitefly Symposium,* 5-9 October, 2004. Cavtat, Crotia. pp.58.

Goyal, C.P. and Brahma, B.C. 2001. A ray of hope against Parthenium in Rajaji National Park. *Indian Forester*, **127**(4): 409–414.

Gunaseelan, V.N. 1987. Parthenium as an additive with cattle manure in bilgas production. *Biological Wastes*, **21**(3): 195–202.

Hakoo, M.L. 1963. A diploid Parthenium in Jammu. *Current Science,* **32**: 273.

Herz, W., Watanabe, H. and Mivazaki, M. 1959. *Journal of American Chemical Society,* **81**: 60–88.

Herz, W., Watanabe, H., Mivazaki, M. and Kishida, Y. 1962. The structures of parthenin and ambrosin. *Journal of American Chemical Society,* **84**: 2601–2610.

Hiremath, I.G. and Y.J. 1997. Parthenium as a source of pesticide. *In*: *Proceedings of the first Intenational Conference on Parthenium Management* held at the University of Agricultural Sciences, Dharwad, Karnataka, India pp. 86–89.

Holm. L.G., Plucknett, D.L., Panchu, J.V. and Herberger, J.P. 1977. *The World's worst weeds.* Honolulu University Press. pp. 109.

Hosmani, M.M. and Prabhakar Setty, T.K., 1973. *Parthenium hysterophorus* Linn. A new weed in Karnataka. *Current Research,* **2**: 93–95.

Jayachandra, 1971. Parthenium weed in Mysore state and its control. *Curr. Sci.* **40**(21): 568–569.

Jayanth, K.P. Visalakshy, G.P.N., Ghosh, S.K. and Chaudhary, M. 1997. Feasibility of biological control of *Parthenium hysterophorus* by *Zygogramma biocolorata* in the light of the controversy due to its feeding on sunflower. *In*: *First International Conference on Parthenium Management* held at Dharwad (Karnataka), 6-8 October, 1997, (*eds.* M. Mahadevappa and V.C. Patil), **I**: 45–51.

Jayanth, K.P. and Geetha Bali. 1993. Temperature tolerance of *Zygogramma bicolorata* introduced for biological control of *Parthenium hysterophorus* (Asteraceae). *Indian Journal of Entomological Research,* **17**: 27–34.

Jayanth, K.P. and Geetha Bali, 1994 Biological control of Parthenium by the beetle *Zygogramma bicolorata* in India. *FAO Plant Protection Bulletin,* **42**: 207–213.

Jayanth, K.P. and Nagarkatti, S. 1987 Investigation on the host specificity and damage potential of *Zygogramma bicolorata* Pallister (Coleoptera:

Chrysomelidae). *Journal of Entomological Research,* **19**(2): 183–185.

Jeevan Rao and shantaram, M.V. 1997 Parthenium menance is soils treated with urban solid wastes. *In*: *Proceedings of the First International Conference on Parthenium Mangement* held at the University of Agricultural Sciences, Dharwad, Karnataka, India. pp. 20–21.

Joseph, K., Manissery, B.S., Narayana Gowda, Shankar, K.M. and Varghese, T.J., 1988. Growth of common carp (*Cyprinus carpio* var. *communis*) fed on feed incorporated with *Cassia tora* leaves. *Indian Journal of Animal Sciences* **58**(6): 737–740.

Kanchan, S.D. 1975. Growth inhibitors from *Parthenium hysterophorus* Linn. *Current Science,* **44**: 358–359.

Kanchan, S.D. 1975a. *Proceedings of Indian Science Congress. Abstract*, III p. 72.

Kanchan, S.D. and Jayachandra. 1976a, Growth inhibitors from *Parthenium hysterophorus* L. *Current Science,* **44**: 358–359.

Kanchan, S.D. and Jayachandra, 1979a. Alelopathic effects of *Parthenium hysterophorus* L. I. Exudation of inhibitors through roots. *Plant Soil,* **53**: 27–35.

Kanchan, S.D. and Jayachandra, 1979b. Alelopathic effects of *Parthenium hysterophorus* L. II. Inhibitory effect of weed residue. *Plant Soil*, **53**: 37–47.

Kanchan, S.D. and Jayachandra, 1980. Allelopathic effects of *Parthenium hysterophorus* L. III. Leaching of inhibitors from aerial vegetative parts. *Plant Soil,* **55:** 61–66.

Kandasamy O.S. and Sankaran, S. 1997. Biological suppression of Parthenium using competitive crops and plants. *In*: *Proceedings of the first International Conference on Parthenium Management* held at the University of Agricultural Sciences, Dharwad, Karnataka, India. pp. 33–36.

Kandhane, D.L., Jangde, C.R. Sadekar, R.D. and Joshirao, M.K. 1992. *Parthenium* toxicity in buffalo calves. *J. Soils Crops.* **21**: 69–71.

Kannabiram 1975. Parthenium, an irritating but promising weed. *Science Reporter,* **12**: 37–38.

Kartar Singh and Shivpuri, D.N. 1971. Studies in yet unknown allergenic pollens of Delhi state metropolitan: Botanical aspect *Indian Journal of Medical Research,* **59**: 1397–1410.

Kasasian, L. and Seeyave, J. 1969. Control of Parthenium (White top). *Annual reports* of the Herbicides Section. Oct. 1967 – Aug. 1968. University of West Indies, Jamaica, *PANS* **15**: 381–398.

Kasliwal, R.M., Bhargava, J.P. and Sogani, I.C. 1961 Value of skin tests in allergy. *Journal of Association Physicians, India* pp. 635–650.

Kaurav, L.P., Chile, A. and Bhan, V.M. 1997 Effect of marigold (*Tegetes papula* Linn.) population on the growth and survival of *Parthenium hysterophorus* L. *In*: *Proceedings of the First International Conference on Parthenium Management* held at the University of Agricultural Sciences, Dharwad, Karnataka, India. pp. 39–40.

Kaurav, L.P., Chile, A. and Bhan,V.M. 1997. Evaluation of *Fusarium pallidoroseum for the bio-control of Parthenium hysteropphorus. Proc. First International Conference on Parthenium Management,* Vol. II. Mahadevappa, M. and Patil, V.C. (*eds.*), 6-8 October, 1997. University of Agricultural Sciences, Dharwad, India. pp. 70–74.

Kelkar, G.R., 1970. Paper presented at symposium on problems caused by *Parthenium hysterophorus* in Maharashtra region, India. Marathi Vignayan Parishad, Poona. July 27, 1968 (*ed.*), G.S. Chandras and V.D. Vartak, *PANS* **16**: 212–214.

Khan, I.S. 1924. The botony of south-west Texas with reference to high fever and asthama. *JAMA* **82**: 871.

Khan, I.S. and Grothaus, E.M. 1936 *Parthenium hysterophorus* antigenic properties, respiratory and cutaneous. *Texas State Journal of Medicine,* **32:** 284.

Khosla, S.N. and Sobti, S.N. 1979. Parthenium – A national health hazard, its control and utility – A review. *Pesticides,* **13**: 121–127.

Kohli, R.K., Kalia, P.K. and Saxena, D.B. 1997. Phytotoxic effects of Parthenium, an allelochemic from *Parthenium hysterophorus* L. on *Ageratum conyzoides* L. – an obnoxious weed. *In*: *Proceedings of the First International Conference on Parthenium Management* held at the University of Agricultural Sciences, Dharwad, Karnataka, India, pp. 140–141.

Kologi, P.D., Kologi, S.D. and Kologi, N.P. 1997. Dermatological hazards of *Parthenium* in human beings. *In*: *First International Conference on Parthenium Management* held at Dharwad (Karnataka), 6-8 October, 1997 (*eds*. M. Mahadevappa and V.C. Patil), **I**: 18–19.

Krishna Murthy, K., Ramachandra Prasad, T.V., Muniyappa, T.V. and Venkata Rao, B.V. 1977. Parthenium, a new pernicious weed in India. University of Agricultural Sciences, *Technical Series*, No. 17, pp. 66.

Krishna Murthy, K., Ramachandra Prasad, T.V. and Muniyappa, T.V. 1976a. Ecology and control of Parthenium. *Proc. Seminar on 'Parthenium a positive danger'*, West End Hotel, Bangalore. Sept. 4, 1976, Organized by Bangalore International cities relationship organization and University of Agril. Sciences, Bangalore pp. 8–10.

Krishna Murthy, K., Ramachandra Prasad, T.V. and Muniyappa, T.V. 1976b. Studies on the control of Parthenium. *Madras Agriculture Journal,* **63**(8-10): 477–480.

Krishna Murthy, K., Ramachandra Prasad, T.V., Muniyappa, T.V. and Venkata Rao, B.V. 1977. Parthenium, a new pernicious weed in India. University of Agricultural Sciences, Technical series, No. **17**, pp. 66.

Kumar, S. and Rohatgi, N. 1999. The role of invasive weeds in changing floristic diversity. *Ann. Forestry,* **7**: 147, 150.

Ladwa, H.R. and Patil, R.M. 1961 Composite of Dharwad and its vicinity. *Journal Bombay Natural History Society,* **58**: 68–70.

Lakshmi Rajan, 1973. Growth inhibitor(s) from *Parthenium hysterophorus* Linn. *Current Science,* **42**: 729–730.

Lonkar, A. and Jog, M.K. 1972. Epidemic contact dermatitis from *Parthenium hysterophorus. Contact Dermatitis Newsletter* (London), Bo. **11**: 291.

Lonkar, A., Mitchell, J.C. and Calman, C.B. 1974. Contact dermatitis from *Parthenium hysterophorus. Transcripts of St. Johns Dermatological Society,* **60**: 43–53.

Mackoff, S. and Dahl, A.O. 1951. A botanical consideration of the weed oleoresin problem. *Minn.* Med., **34**: 1169.

Mahadevappa, M. and Patil. V.C. 1997. University of Agricultural Sciences, Dharwad, India. pp. 55–62.

Mahadevappa, M. and Joshi, S.S. 1985. Biological control of Parthenium, *Seed Tech. News.,* **15**: 21–22.

Mahadevappa, M., Kulkarni, R.S., Munegowda, M.K. and Parmeshwar, N.S. 1990. Establishment of *Parthenium hysterophorus* under different soil and climatic conditions of Karnataka. *Indian Journal of Weed Science* **21:** 65–68.

Mahadevappa, M. and Kulkarni, R.S. 1991 Control of *Parthenium hysterophorus* L. through biological agents. International Conference on "Biological Control I Tropical Agriculture" August 26-30, 1991 at Kauala Lampur, Malasia.

Mahadevappa, M. 1996. *Parthenium.* Prasaranga, University of Mysore, Mysore, India. pp. 60.

Mahadevappa, M. 1997. Ecology, distribution, menace and management of Parthenium. *In*: *First International Conference on Parthenium management* held at Dharwad, (Karnataka), October 6-8, 1997, **I**: 1–12.

Mahadevappa, M. and Ramaiah, H. 1988. Pattern of replacement of *Parthenium hysterophorus* plants by *Cassia serecia* in waste lands. *Indian Journal of Weed Science,* **20**(4): 83–85.

Mahadevappa, M. 1999. *Parthenium and its management.* University of Agricultural Sciences, Dharwad (Karnataka). pp. 132.

Mahadevappa, M. and Gautam, R.D. 2006. Introduction of *Parthenium hysterophorus* L.: Status and Management. *Indian J. Plant Genet. Resour.* **19**(3): 453–459.

Mahesha, H.M., Rajashekhargouda, R. and Rayer, S.G. 1999. Effect of aqueous extracts of few botanicals with special reference to weeds on *Bombyx mori* L. The proceeding of the XVIIth International Sericultural Commission Congress held at Cairo-Egypt during 12-16th October, 1999. pp. 114–121.

Maheshwari, J.K. 1966. *Parthenium hysterophorus. Current Science*, **35**: 181–183.

Maheshwari, J.K. 1966a. *Illustrations to the flora of Delhi,* New Delhi, pp. 17.

Maheshwari, J.K., 1968. Parthenium weed in Madhya Pradesh. *Current Science*, **37**: 326–327.

Maheshwari. J.K. and Pandey, R.S. 1973. Parthenium weed in Bihar State. *Current Science*, **42**: 733.

Maiti, G.S. 1983. An untold study on the occurrence of *Parthenium hysterophorus* L. in India. *India J. Forestry*, **6:** 328–329.

Mamatha, M. and Mahadevappa, M. 1988. Biological survey in relation to Parthenium control. *Advances in Plant Science,* **1**(2): 233–238.

Mamatha, M. and Mahadevappa, M. 1992. Biological survey in relation to Parthenium control. *Advances in Plant Science, Special Issue*, pp. 426–429.

Mathur, S.K. and Muniyappa, V. 1989. *Mimiographed Report* – University of Agricultural Sciences, Bangalore, India. pp. 33.

Mitchell, J.C. and Dupuis, G., 1971. Allergic contact dermatitis from sesquiterpenoids of the compositae family of plants. *British Journal of Dermatology,* **89**: 139.

Mitchell, J.C., Dupuis, G. and Gissman, T.A. 1972. Allergic contact dermatitis from sesquiterpenodis of palnts. Additional allergenic sesquiterpene lactones. *British Journal of Dermatology*, **87**: 235.

McClay, A.S. 1980. *Mimeographed Report.* Commonwealth Institute of Biological Control, Monterry, Mexico. pp. 10.

Muniyappa, T.V. 1980. Biology and control of *Parthenium hysterophorus* L. and its allelopathic effect on field crops. M.Sc. thesis, University of Agricultural Sciences, Bangalore, India. pp. 118.

Navie, S.C., McFadyen, R.E., Panetta, F.D. and Adkins, S.W. 1996. The biology of Australian weeds. *Parthenium hysterophorus* L. *Plant Protection Quarterly* **11**: 76–88.

Navie, S.C., Panetta, F.D. McFadyen, R.E. and Adkins, S.W. 1998a. Behaviour of buried and surface sown seeds of *Parthenium hysterophorus* L. *Weed Research* **38**: 335–341.

Navie, S.C., Priest, T.E., McFadyen, R.E. and Adkins, S.W. 1988b. Efficacy of the stem-galling moth *Epiblema strenuana* Walk. (Lepidoptera : Tortricidae) as a biological control agent for ragweed Parthenium (*Parthenium hysterophorus* L.). *Biological Control* **13**: 1–8.

Navie, S.C., McFadyen, R.E., Panetta, F.D. and Adkins, S.W. 1988c. *Parthenium hysterophorus* L. *In* : *The Biology of Australian Weeds* (*eds.* Victoria, R.G. and F.J. Richardson). pp. 157–176.

Nagaraja, A., Yogeesha, H. S. and Shekhargouda, M., 1997. Effect of Parthenium extracts on seed germination and vigour of different crops species. *In*: *Proceedings of the First International Conference on Parthenium Management* held at the University of Agricultural Sciences, Dharwad, Karnataka, India. pp. 154–156.

Narasimhan, T.R., Ananth, M., Narayanaswamy, M., Rajendra Babu, M., Mangala, A. and Subba Rao, P.V. 1977. Toxicity of *Parthenium hysterophorus* L. to cattle and buffaloes. *Experientia* **33**: 1358–1359.

Ogden, H.D. 1975. Diagnosis and treatment of Parthenium dermatitis. Journal of Louisiana Medical Society, **109**: 378.

Opler, Morris, E. 1946. *Childhood and youth in Jicarilla Apache society*. Publications of the Frederick Webb Hodge Anniversary Fund (Vol. 5). Los Angeles: The Southwest Museum Administrator of the Fund.

Ovies, J. and Larringa, L. 1988. Transmission of *Xanthamonas campestris pv. Phaseoli* by a wild host. *Ciencia Y Tencica en la Agricultura Protection de Plants* **11:** 23–30.

Pandey. P.K. and Saini, S.K. 2002. Parthenium: Threat to biological diversity of natural forest ecosystem in Madhya Pradesh. *Vaniki Sandesh*, **26**(2): 21–29.

Parker, A. 1990. Biological control of Parthenium weed using two rust fungi. Proc. VIII International Symposium on Biological Control of Weeds CAB International Istitute of Biological Control, Silwood Park, Ascot. Berks, UK. pp. 531–537.

Parker, A., Holden, A.N.G. and Tomely, A.J. 1994. Host specificity testing and assessment of the rust, *Puccinia abrupta* var *partheniicola* as a biological control agent of Parthenium weed (*Parthennium hysterophorus*). *Plant Pathology* **43**: 1–16.

Parmelee, J.A. 1967. The autoecious species of *Puccinia* on Heliantheae in North America. *Canadian J. Botany* **45**: 2267–2328.

Patel, V.S, Chitra, V., Prasanna, P.L. and Krishnaraju, V. 2008. Hypoglycemic effect of aqueous extract of *Parthenium hysterophorus* L. in normal and alloxan induced diabetic rats. *Indian J. Pharmacol* **40**: 183–185.

Patil R.R. Mahadevappa, M., Mahesha, H.M. and Patil, V.C. 1997a. Phagostimulant effect of Parthenium on mulberry silkworm (*Bombyx mori* L.). *In*: *Proceedings of the First International Conference on Parthenium Management* held at the University of Agricultural Sciences, Dharwad, Karnataka, India. pp. 81–85.

Patil, S. A., Jatti, P.D., Chetti, M.B. and Hiremath, S.M. 1997c. A survey of Parthenium in Bijapur district – A case study. *In*: *Proceedings of the First International Conference on Parthenium Management* held at the

University of Agricultural Sciences, Dharwad, Karnataka, India pp. 20–23.

Prasada Rao, R.D.V.J., Reddy, A.S. Chander Rao, S. Varaprasad, K.S. Thrirumala Devi, Nagaraju, K. Muniyappa, V. and Reddy, D.V.R. 2000. Tobacco streak virus as causal agent of sunflower necrosis disease in India. *J. Oilseeds Reasearch* **17**: 400–401.

Prasada Rao, R.D.V.J., Reddy, A.S., Reddy, S.V., Thirumala Devi, K., Chander Rao, S., Manoj Kumar, V., Subramaniam, K., Tellamanda Reddy, T., Nigam, S.N. and Reddy, D.V.R. 2003. The host range of tobacco streak virus in India and transmission by thrips. Annals of *Applied Biology* **142**: 365–368.

Puttaswamy, Devaiah, M.C. and Rangaswamy, M.R. 1976. A new host record of the vegetable mite, *Tetranychus cucurbitae* Rahman and Sapra. *Current Science,* **45**: 118.

Rice, E.L. 1984. *Allelopathy*. Academic Press, New York.

Rajwar, G.S., Haigh, M.J. Krecek, J. and Kilmartin, M.P. 1998. Changes in plant diversity and related problems for environmental management in the headwaters of the Garhwal Himalaya. **In:** *Proceedings of Headwaters: Water Resources and Soil Conservation,*'98, Fourth International Conference on Headwater Control. Merano, Italy. pp. 335–343.

Ramaswami, P.P. 1997. Potential uses of *Parthenium. In*: *First International Conference on Parthneium Management* held at Dharwad (Karnataka), 6-8 October, 1997 (*eds*. M. Mahadevappa and V.C. Patil), **I**: 77–80.

Ranade, S. 1970. Paper presented at "Symposium on Problems Caused by *Parthenium hysterophorus* in Maharashtra Region, India. Marathi Vignayan Parishad, Poona, July 27, 1968 (*eds*.). G.S. Chandras and V.D. Vartak, *PANS* **16**: 12–14.

Ranade, S. 1975. Results of newly synthesized vaccine in case of congress-grass eczema in India. Paper read at "Aswini" Naval base hospital, Bombay.

Ranade, S. 1976. Results of newly synthesized vaccine in

cases of Parthenium eczema in India. *Proceedings of Seminar on Parthenium – A positive* danger. West End Hotel, Bangalore, Sept. 4, 1976, Organized by Bangalore International cities relationship organization and University of Agricultural Sciences, Bangalore. pp. 15–16.

Rao R.S. 1956 Parthenium a new record for India. *Journal of Bombay Natural History Society,* **54**: 218–220.

Ramadevi, O.K., Theagarajan, K. S., Gavikumar, G. and Raja Mukhurishnan, 1997. Suppression of Parthenium using extracts from eucalyptus tree. *In*: *Proceedings of the First International Conference on Parthenium Management* held at the University of Agricultural Sciences, Dharwad, Karnataka, India. pp. 36–38.

Reed, C.F. 1964. *Phytologica* **10**: 338.

Roberts, J. 1967. The carrot weed problem: Poona area. Report to Dr. K. Subramanyam, Director, Botanical Survey of India, Calcutta (Quoted by Lokar Mitchell, Calnan, 1974). *Transcripts of St. Johns' Hospital Dermatological Society*, **60**(1): 43–53.

Roxburgh, W. 1914. *Hortus Bengalensis*, Mison Press Calcutta.

Satyaprasad, K. and Usharani, P. 1981. Occurrence of powdery mildew on Parthenium caused by *Oidium parthenii. Current Science* **24**: 1081–1082.

Santapau, H. 1967 The flora of Khandala on the western ghats of India. *Record of Botanical Survey of India,* **16**: 1–372 (Revised edition).

Santha, A., Reddy, G.L.N. and Narayanan, A. 1997. Effect of Parthenium leaf extracts on crop used germination. *In*: *Proceedings of the First International Conference on Parthenium Management* held at the University of Agricultural Sciences, Dharwad, Karnataka, India. pp. 157–159.

Seier, M.K., Harvey, J.L. Romero, A. 1997. Satety testing of the rust *Puccinia melampodii* as a potential biocontrol agent of *Parthenium hysterophorus.* Proc. *First International Conference on Parthenium Management,* Vol. II. Mahadevappa. M. and Patil, V.C.

(*eds.*), 6-8 October, 1997. University of Agricultural Sciences, Dharwad, India. pp. 63–69.

Sethuraman, G., Bansal, A., Sharma, V.K. and Verma, K.K. 2008. Seborrhoeic pattern of Parthenium dermatitis. *Contact Dermatitis*. **58**: 372–374.

Sharma, V.K., Sethuraman, G., Garg, T, Verma, K.K. and Raman, M. 2004. Patch testing with the Indian standard series in New Delhi. *Contact Dermatitis* **51**: 319–321.

Shelmire, B. 1939. Contact dermatitis from vegetation. *South Medical Journal*, **33**: 338.

Shelmire, B. 1939. Contact dermatitis from weeds: Patch testing with their oleoresins. *JAMA* **113**: 1085.

Shelmire, B. 1940. Contact dermatitis from vegetation. *South Medical Journal*, **33**: 338.

Shukla, K.S. and Singh, S.P. 1999. *Research in composite weed-achievement during last 50 years*. Forest Research Institute, Dehradun, India. pp. 33.

Singh, A.B., Menon, M.P.S. and Shivpuri, D.N. 1974. Preliminary report on the allergenicity of various parts of *Parthenium hysterophorus* aspects. *Allergic Appl. Immunology*, **7**: 47–51.

Singh, N.P. 1983. Potential biological control of *Parthenium hysterophorus* L. *Current Science* **52**: 644.

Singla, R.K. 1992. Can Parthenium be put to use? The *Tribune*, **112**(294): 6.

Son, T.T.N. and Ramaswami, P.P. 1997. Bioconversion of Parthenium for sustainable agriculture. *In*: *Proceedings of the first International Conference on Parthenium Management* held at the University of Agricultural sciences, Dharwad, Karnataka, India. pp. 119–123.

Sreenivasa, M.N. and Majjigudda, I.M. 1997. Biogas conversion from few weeds. *In*: *Proceedings of the First International Conference on Parthenium Management* held at the University of Agricultural Sciences, Dharwad, Karnataka, India. pp. 124–125.

Sridhar, S. 1991. A cure no more, under attack the sunflower crop. *Frontline,* Madras, Nov. 9-22. pp. 100.

Subba Rao, P.V. Mangala, A., Seetharamaiah, Subba Rao.B.S. and Prakash, K.M. 1976 *Parthenium* - An allergic weed. *Proc. Seminar on 'Parthneium - A positive danger'* West End Hotel, Banglaore, Sept. 4, 1976, Organised by Bangalore International cities relationship organization and Univ. of Agril. Sciences, Banglaore. pp. 17–18.

Sukhada, D.K. 1975. Growth inhibitors from *Parthenium hysterophorus* L. *Current Science* **44**: 358–359.

Sundara Rajulu, G. and Gowri. M. 1976 Biological control of the poisonous weed *Parthenium. Proc. of seminar on Parthenium - A positive danger.* West End Hotel, Bangalore. Sept. 4, 1976, Organized by Bangalore International cities relationship organization and the University of "Agril. Sciences, Bangalore. pp. 22–26.

Suresh Chandra. 1973. Parthenium - A new dermatitis weed in Bihar. *Proceedings of Sixtieth Indian Science Congress,* Chandigarh III p. 375.

Suresh, P.V., Gupta, D., Behera, D. and Jindal, S.K. 1994. Bronchial provocation with Parthenium, Pollen extract in bronchial asthma. *Indian J. Chest Disease Allied Sci.,* **36**: 104.

Sushilkumar and Bhan, V.M. 1988. Imported Mexican beetle *Zygogramma biocolorata* involved in sunflower feeding controversy emerged out as a safe bioagents against *Parthenium hysterophorus* in India. *8th Biennial Conference of Indian Society of Weed Science,* 5-7 February, 1999. At Banaras (UP), India. pp. 135.

Sushil Kumar and Saraswat, V.N. 2001. Integrated management: The only solution to suppress *Parthenium hysterophorus. In*: *"Alien Weeds in Moist Tropical Zones: Banes and Benefits"* (*eds.* K.V. Sankaran, S.T. Murphy and H.C. Evans), Workshop Proceedings, 2-4 November, 1999, Kerala, India, Kerala Forest Research Institute, India and CABI Biooscience, UK Centre (As-cot), UK. pp. 150–168.

Taneja, K., Jain, P.P. and Naithani, S. 1997. Utilization of *Parthenium hysterophorus* (Gajar grass) and *Circium sp.* as paper raw material. *World Weeds* **4**: 21–23.

Thimmaiah, A. and Bhatnagar, R.K. 1997. Vermitechnology for the Utilization of *Parthenium hysterophorus* L. *In*: *Proceeding of the First International Conference on Parthenium Management* held at the University of Agricultural Sciences, Dharwad, Karnataka, India. pp. 128–129.

Tiwari, J.P. and Bisen, G.R. 1982. Growth analysis of *Parthenium hysterophorus* L. *Indian, Journal of Weed Science* **14**: 49–53.

Tiwari, J.P. and Bisen, G.R. 1984, Ecology of *Parthenium hysterophorus Indian. Journal of Weed Science* **16**: 203–206.

Tomley, A.J. and Evans, H.C. 2004. Establishment and preliminary impact of the Madagascaru rust, *Maravalia crytostegiae* grandiflora (Asclepiadaceae) in Queensland, Australia. *Plant Pathol.* **53**: 475–484.

Towers, G.H.N., Mitchell, J.C., Redrigurz, E., Bennett, F.D. and Subba Rao. P.V. 1977. Biology and chemistriy of *Parthenium hysterophorus* LK. – A problem weed in India. *J. Sci. Ind. Res.* **36**: 672–684.

Tripathi, B., Barla, A. and Singh, C.M. 1991. Carrot weed – *Parthenium hysterophorus* L. A review of the problem and strategy for its control in India. *Indian Journal of Weed Science* **23**: 61–71.

Uphof, J.C.T. 1959. *Dictionary of economic plants*. N.Y. Engelmann, Hafner Publishing co., New York. pp. 267.

Vaid, K.M. and Gand Naithani, H.B. 1970. *Parthenium hysterophorus* Linn. A new record for the North Western Himalaya. *Indian Forester* **96**: 791–792.

Vartak, V.D. 1968. Weed that threatens crop and grass lands in Maharashtra. *Indian Farming* **18**(1): 23–24.

Verma, K. Mahesh, R. Srivastava, P. and Ramam, M. 2007. Comparison of effectiveness and safety of azathioprine versus betamethasone for the treatment of Parthenium dermatitis. *Brit. J. Dermatol* **157**: 9.

Verma, K.K. 2007. Type I hypersensitivity to *Parthenium hysterophorus* in patients with Parthenium dermatitis. *Indian J. Dermatol. Venereol. Leprol.* **73**: 265.

Verma, K.K., Bansal, A. and Sethuraman, G. 2006 Parthenium dermatitis treated with azathioprine weekly pulse doses. *Ind. J. Dermatol. Venereol. Leprol.* **72**: 24–27.

Verma, K.K. Manchanda, Y. and Dwivedi, S.N. 2004 Failure of titre of contact hypersensitivity to correlate with clinical severity and therapeutic response in contact dermatitis caused by *Parthenium*. *Ind. J. Dermatol. Venereol. Leprol.* **70**: 210–213.

Verma, K.K., Sirka, C.S., Ramam, M. and Sharma, V.K. 2002. Parthenium dermatitis presenting as photosensitive lichenoid eruption: A new clinical variant. *Contact Dermatitis* **46**: 286–289.

Verma, K.K., Sirka, C.S. and Ramam, M. 2001. Contact dermatitis due to plants – Challenges and prospects. *Ind. Pract.* **54**: 791–796.

Verma, K.K., Manchanda, Y. and Pasricha, J.S. 2000. Azathioprine as a corticosteroid sparing agent for the treatment of dermatitis caused by the weed Parthenium. *Acta Dermatol. Venereol.* **80**: 31–32.

Whittaker, R.H. and Feeny, P.P. 1971. Allelochemics: Chemical interactions between species. *Science*, **171**: 757.

Woodhouse, R.P. 1971. *Hayfever plants*. 2nd edn. New York, Hafner Publishing Co. pp. 154.

Additional References

Anon. 1997. Parthenium weed (*Parthenium hysterophorus*) Pestfact PP2. Qld Gov., Brisbane.

Holman, D.J and Dale, I.J. 1981. Parthenium weed threatens Bowen Shire. *Qld Ag. J.*, **107**: 57–60.

Navie, S.C., Panetta, F.D., McFadyen, R.E., and Adkins, S.W. Longevity of buried and surface-lying seeds of Parthenium weed (in press).

Parsons, W.T. and Cuthbertson, E.G. 1992. Noxious Weeds of Australia. Inkata Press, Melbourne. pp. 692.

Stuessy, T.F. 1977. Heliantheae, a Systematic Review. *In*:

The Biology and Chemistry of the Compositae. Vol. II. *eds*. V.H. Heywood, J.B. Harbone and B.L. Turner, Acad. Press, London.

Towers, G.H.N. and Subba Rao, P.V. 1992. Impact of the pan-tropical weed *Parthenium hysterophorus* L. on human affairs. *In*: *Proc.1st Int. Weed Control Congr. Melbourne* (*ed*. R.G. Richardson) Melbourne, Australis. pp.134–138.

Tudor, G.D., Ford, A.L., Armstrong, T.R. and Bromage, E.K. 1982. Taints in meat from sheep grazing Parthenium weed. *Aust. J. Exp. Agric. Husb.* **22**: 43–46.

Chippendale, J.F. and Panetta, F.D. 1994. The cost of Parthenium weed to the Queensland cattle industry. *Plant Protection Quarterly* **9**: 73–76.

Navie, S.C., McFadyen, R.E., Panetta, F.D. and Adkins, S.W. 1996. A comparison of the growth and phenology of two introduced biotypes of *Parthenium hysterophorus*. *In*: Proc. 11th Aust. Weeds Conf., R.C.H. Shepherd (*ed.*) Weed Sci. Soc. Victoria, Frankston. pp. 313–316.

Adkins, S.W., Navie S.C. and McFadyen R.E. 1996. Control of Parthenium Weed (*Parthenium hysterophorus* L.): A Centre for Tropical Pest Management Team Effort pp. 573–578. *In*: Proc. 11th Aust. Weeds Conf., R.C.H. Shepherd (*ed.*) Weed Sci. Soc. Victoria, Frankston.

Navie, S.C., McFadyen, R.E., Panetta, F.D. and Adkins, S.W. 1996. The biology of Australian weeds. *Parthenium hysterophorus* L. *Plant Prot. Quart.* **11**: 76–88.

Dhileepan, K., Madigan, B., Vitelli, M., McFadyen, R.E., Webster, K. and Trevino, M. 1996. A new initiative in the biological control of Parthenium. *In*: Proc. 11th Aust. Weeds Conf., R.C.H. Shepherd (*ed.*) Weed Sci. Soc. Victoria, Frankston. pp. 309–312.

McClay, A.S., Palmer, W.A., Bennett, F.D. and Pullen, K.R. 1995. Phytophagous arthropods associated with *Parthenium hysterophorus* (Asteraceae) in North America. *Environ. Entomol.* **24**: 796–809.

McFadyen, R.E. and Cruttwell. 1992. Biological control against Parthenium weed in Australia. *Crop Protection*, **11**: 400–407.

SUBJECT INDEX

D

E

F

INTERNATIONAL CONFERENCE ON PARTHENIUM MANAGEMENT